Lecture Notes
in Business Information Processing   **360**

Series Editors

Wil van der Aalst
RWTH Aachen University, Aachen, Germany
John Mylopoulos
University of Trento, Trento, Italy
Michael Rosemann
Queensland University of Technology, Brisbane, QLD, Australia
Michael J. Shaw
University of Illinois, Urbana-Champaign, IL, USA
Clemens Szyperski
Microsoft Research, Redmond, WA, USA

More information about this series at http://www.springer.com/series/7911

Thomas Hildebrandt · Boudewijn F. van Dongen ·
Maximilian Röglinger · Jan Mendling (Eds.)

Business Process Management Forum

BPM Forum 2019
Vienna, Austria, September 1–6, 2019
Proceedings

 Springer

Editors
Thomas Hildebrandt [iD]
University of Copenhagen
Copenhagen, Denmark

Boudewijn F. van Dongen [iD]
Eindhoven University of Technology
Eindhoven, The Netherlands

Maximilian Röglinger [iD]
University of Bayreuth
Bayreuth, Germany

Jan Mendling [iD]
Vienna University of Economics
and Business
Vienna, Austria

ISSN 1865-1348   ISSN 1865-1356 (electronic)
Lecture Notes in Business Information Processing
ISBN 978-3-030-26642-4   ISBN 978-3-030-26643-1   (eBook)
https://doi.org/10.1007/978-3-030-26643-1

This Springer imprint is published by the registered company Springer Nature Switzerland AG
The registered company address is: Gewerbestrasse 11, 6330 Cham, Switzerland

Preface

This volume contains the papers presented at the BPM Forum of the 17th International Conference on Business Process Management (BPM 2019). The conference provided a forum for researchers and practitioners in the broad and diverse field of BPM. To accommodate for the diversity of the field, the BPM conference hosted tracks for foundations, engineering, and management. The conference was held in Vienna, Austria, during September 1–6, 2019.

Since its introduction, the aim of the BPM Forum has been to host innovative research that has high potential to stimulate discussion, but does not quite meet the rigorous quality criteria for the main conference. The papers selected for the forum showcase fresh ideas from exciting and emerging topics in business process management. We received a total of 157 submissions, from which we took 115 into review. Each submission was reviewed by at least three Program Committee (PC) member and one senior PC member. In all, 23 papers were accepted for the main conference. In addition, we invited 17 innovative papers to the BPM Forum, out of which 13 accepted our invitation.

We thank the colleagues involved in the organization of the conference, especially the members of the PCs and the Organizing Committee. We also thank the Platinum sponsor Signavio, the Gold sponsors Austrian Center for Digital Production, Bizagi, Camunda, Celonis, FireStart, Process4.biz, the Silver sponsors Heflo, JIT, Minit, Papyrus Software, Phactum, and the Bronze sponsors ConSense, DCR, and TIM Solutions, as well as Springer and Gesellschaft für Prozessmanagement for their support. We also thank WU Vienna and the University of Vienna for their enormous and high-quality support. Finally, we thank the Organizing Committee and the local Organizing Committee, namely, Martin Beno, Katharina Distelbacher-Kollmann, Ilse Dietlinde Kondert, Roman Franz, Alexandra Hager, Prabh Jit, and Doris Wyk.

September 2019

Thomas Hildebrandt
Boudewijn van Dongen
Maximilian Röglinger
Jan Mendling

Organization

The 17th International Conference on Business Process Management (BPM 2019) was organized by the Vienna University of Economics and Business (WU Vienna) and the University of Vienna, and took place in Vienna, Austria.

Steering Committee

Mathias Weske (Chair)	HPI, University of Potsdam, Germany
Boualem Benatallah	University of New South Wales, Australia
Jörg Desel	Fernuniversität in Hagen, Germany
Schahram Dustdar	TU Wien, Austria (until 2018)
Marlon Dumas	University of Tartu, Estonia
Wil van der Aalst	RWTH Aachen University, Germany
Michael zur Muehlen	Stevens Institute of Technology, USA (until 2018)
Barbara Weber	University of St. Gallen, Switzerland
Stefanie Rinderle-Ma	University of Vienna, Austria
Manfred Reichert	University of Ulm, Germany
Jan Mendling	WU Vienna, Austria

Executive Committee

General Chairs

Jan Mendling	WU Vienna, Austria
Stefanie Rinderle-Ma	University of Vienna, Austria

Main Conference Program Chairs

Thomas Hildebrandt	University of Copenhagen, Denmark (Track I)
Boudewijn van Dongen	Eindhoven University of Technology, The Netherlands (Track II)
Maximilian Röglinger	University of Bayreuth, Germany (Track III)
Jan Mendling	WU Vienna, Austria (Consolidation)

Blockchain Forum Chairs

Claudio Di Ciccio	WU Vienna, Austria
Luciano García-Bañuelos	Tecnológico de Monterrey, Mexico
Richard Hull	IBM Research, USA
Mark Staples	Data61, CSIRO, Australia

Central Eastern European Forum Chairs

Renata Gabryelczyk	University of Warsaw, Poland
Andrea Kő	Corvinus University of Budapest, Hungary
Tomislav Hernaus	University of Zagreb, Croatia
Mojca Indihar Štemberger	University of Ljubljana, Slovenia

Industry Track Chairs

Jan vom Brocke	University of Liechtenstein, Liechtenstein
Jan Mendling	WU Vienna, Austria
Michael Rosemann	Queensland University of Technology, Australia

Workshop Chairs

Remco Dijkman	Eindhoven University of Technology, The Netherlands
Chiara Di Francescomarino	Fondazione Bruno Kessler-IRST, Italy
Uwe Zdun	University of Vienna, Austria

Demonstration Chairs

Benoît Depaire	Hasselt University, Belgium
Stefan Schulte	TU Wien, Austria
Johannes de Smedt	University of Edinburgh, UK

Tutorial Chairs

Akhil Kumar	Penn State University, USA
Manfred Reichert	University of Ulm, Germany
Pnina Soffer	University of Haifa, Israel

Panel Chairs

Jan Recker	University of Cologne, Germany
Hajo A. Reijers	Utrecht University, The Netherlands

BPM Dissertation Award Chair

Stefanie Rinderle-Ma	University of Vienna, Austria

Event Organization Chair

Monika Hofer-Mozelt	University of Vienna, Austria

Local Organization Chairs

Katharina Disselbacher-Kollmann	WU Vienna, Austria
Roman Franz	WU Vienna, Austria
Ilse Dietlinde Kondert	WU Vienna, Austria

Publicity Chair

Cristina Cabanillas   WU Vienna, Austria

Web and Social Media Chair

Philipp Waibel   WU Vienna, Austria

Proceedings Chair

Claudio Di Ciccio   WU Vienna, Austria

Track I – Foundations

Senior Program Committee

Florian Daniel	Politecnico di Milano, Italy
Jörg Desel	Fernuniversität in Hagen, Germany
Chiara Di Francescomarino	Fondazione Bruno Kessler-IRST, Italy
Dirk Fahland	Eindhoven University of Technology, The Netherlands
Marcello La Rosa	The University of Melbourne, Australia
Fabrizio Maria Maggi	University of Tartu, Estonia
Marco Montali	Free University of Bozen-Bolzano, Italy
John Mylopoulos	University of Toronto, Canada
Manfred Reichert	University of Ulm, Germany
Victor Vianu	University of California San Diego, USA
Hagen Völzer	IBM Research – Zurich, Switzerland
Mathias Weske	HPI, University of Potsdam, Germany

Program Committee

Ahmed Awad	University of Tartu, Estonia
Jan Claes	Ghent University, Belgium
Søren Debois	IT University of Copenhagen, Denmark
Claudio Di Ciccio	WU Vienna, Austria
Rik Eshuis	Eindhoven University of Technology, The Netherlands
Hans-Georg Fill	University of Fribourg, Switzerland
Guido Governatori	Data61, CSIRO, Australia
Gianluigi Greco	University of Calabria, Italy
Richard Hull	IBM Research, USA
Irina Lomazova	National Research University Higher School of Economics, Russia
Andrea Marrella	Sapienza University of Rome, Italy
Oscar Pastor Lopez	Universitat Politècnica de València, Spain
Artem Polyvyanyy	The University of Melbourne, Australia
Wolfgang Reisig	Humboldt-Universität zu Berlin, Germany
Arik Senderovich	University of Toronto, Canada
Tijs Slaats	University of Copenhagen, Denmark

Ernest Teniente   Universitat Politècnica de Catalunya, Spain
Daniele Theseider Dupré   Università del Piemonte Orientale, Italy

Track II – Engineering

Senior Program Committee

Boualem Benatallah   The University of New South Wales, Australia
Josep Carmona   Universitat Politècnica de Catalunya, Spain
Cesare Pautasso   University of Lugano, Switzerland
Hajo A. Reijers   Utrecht University, The Netherlands
Pnina Soffer   University of Haifa, Israel
Wil van der Aalst   RWTH Aachen University, Germany
Ingo Weber   Data61, CSIRO, Australia
Barbara Weber   University of St. Gallen, Switzerland
Matthias Weidlich   Humboldt-Universität zu Berlin, Germany
Lijie Wen   Tsinghua University, China

Program Committee

Marco Aiello   University of Stuttgart, Germany
Amin Beheshti   Macquarie University, Australia
Andrea Burattin   Technical University of Denmark, Denmark
Cristina Cabanillas   WU Vienna, Austria
Fabio Casati   University of Trento, Italy
Massimiliano de Leoni   University of Padua, Italy
Jochen De Weerdt   KU Leuven, Belgium
Remco Dijkman   Eindhoven University of Technology, The Netherlands
Marlon Dumas   University of Tartu, Estonia
Schahram Dustdar   TU Wien, Austria
Gregor Engels   Paderborn University, Germany
Joerg Evermann   Memorial University of Newfoundland, Canada
Walid Gaaloul   Télécom SudParis, France
Avigdor Gal   Technion, Israel
Luciano García-Bañuelos   Tecnológico de Monterrey, Mexico
Chiara Ghidini   Fondazione Bruno Kessler (FBK), Italy
Daniela Grigori   Laboratoire LAMSADE, University Paris-Dauphine, France
Dimka Karastoyanova   University of Groningen, The Netherlands
Christopher Klinkmüller   Data61, CSIRO, Australia
Agnes Koschmider   Karlsruhe Institute of Technology, Germany
Jochen Kuester   FH Bielefeld, Germany
Henrik Leopold   Kühne Logistics University, Germany
Raimundas Matulevičius   University of Tartu, Estonia
Massimo Mecella   Sapienza University of Rome, Italy
Hamid Motahari   Ernst & Young (EY), USA
Jorge Munoz-Gama   Pontificia Universidad Católica de Chile, Chile

Hye-Young Paik	The University of New South Wales, Australia
Luise Pufahl	Hasso Plattner Institute, University of Potsdam, Germany
Manuel Resinas	University of Seville, Spain
Minseok Song	POSTECH, Pohang University of Science and Technology, South Korea
Arthur ter Hofstede	Queensland University of Technology, Australia
Farouk Toumani	Limos, Blaise Pascal University, Clermont-Ferrand, France
Eric Verbeek	Eindhoven University of Technology, The Netherlands
Moe Wynn	Queensland University of Technology, Australia
Bas van Zelst	Fraunhofer Gesellschaft Aachen, Germany

Track III – Management

Senior Program Committee

Jörg Becker	University of Münster, ERCIS, Germany
Adela Del Río Ortega	University of Seville, Spain
Marta Indulska	The University of Queensland, Australia
Susanne Leist	University of Regensburg, Germany
Mikael Lind	Research Institutes of Sweden (RISE)/Chalmers University of Technology, Sweden
Peter Loos	IWi at DFKI, Saarland University, Germany
Jan Recker	University of Cologne, Germany
Michael Rosemann	Queensland University of Technology, Australia
Flavia Santoro	University of the State of Rio de Janeiro, Brazil
Peter Trkman	University of Ljubljana, Slovenia
Amy Van Looy	Ghent University, Belgium
Robert Winter	University of St. Gallen, Switzerland

Program Committee

Wasana Bandara	Queensland University of Technology, Australia
Daniel Beimborn	University of Bamberg, Germany
Daniel Beverungen	Paderborn University, Germany
Alessio Maria Braccini	University of Tuscia, Italy
Michael Fellmann	University of Rostock, Institute for Computer Science, Germany
Peter Fettke	German Research Center for Artificial Intelligence (DFKI) and Saarland University, Germany
Kathrin Figl	University of Innsbruck, Austria
Andreas Gadatsch	Hochschule Bonn-Rhein-Sieg, Germany
Paul Grefen	Eindhoven University of Technology, The Netherlands
Thomas Grisold	University of Liechtenstein, Liechtenstein
Bernd Heinrich	Universität Regensburg, Germany
Mojca Indihar Štemberger	University of Ljubljana, Slovenia

Christian Janiesch	TU Dresden, Germany
Florian Johannsen	University of Applied Sciences Schmalkalden, Germany
John Krogstie	Norwegian University of Science and Technology, Norway
Michael Leyer	University of Rostock, Germany
Alexander Mädche	Karlsruhe Institute of Technology, Germany
Monika Malinova	WU Vienna, Austria
Juergen Moormann	Frankfurt School of Finance and Management, Germany
Michael zur Muehlen	Stevens Institute of Technology, USA
Markus Nüttgens	University of Hamburg, Germany
Sven Overhage	University of Bamberg, Germany
Geert Poels	Ghent University, Belgium
Jens Poeppelbuss	Ruhr-Universität Bochum, Germany
Michael Räckers	University of Münster, ERCIS, Germany
Shazia Sadiq	The University of Queensland, Australia
Bernd Schenk	University of Liechtenstein, Liechtenstein
Werner Schmidt	Technische Hochschule Ingolstadt (THI Business School), Germany
Theresa Schmiedel	University of Applied Sciences and Arts Northwestern Switzerland, Switzerland
Anna Sidorova	University of North Texas, USA
Oktay Türetken	Eindhoven University of Technology, The Netherlands
Irene Vanderfeesten	Eindhoven University of Technology, The Netherlands
Jan Vanthienen	KU Leuven, Belgium
Axel Winkelmann	University of Würzburg, Germany

Additional Reviewers

Abasi-Amefon Affia
Simone Agostinelli
Ivo Benke
Sabrina Blaukopf
Djordje Djurica
Montserrat Estañol
Florian Fahrenbach
Christian Fleig
Lukas-Valentin Herm
Mubashar Iqbal
Samuel Kießling

Fabienne Lambusch
Xavier Oriol
Baris Ozkan
Michael Poppe
Raphael Schilling
Roee Shraga
Ludwig Stage
Peyman Toreini
Jonas Wanner
Bastian Wurm

Contents

Management

Specification

Sketching Process Models by Mining Participant Stories

Ana Ivanchikj[(✉)] and Cesare Pautasso

Software Institute, USI, Lugano, Switzerland
{ana.ivanchikj,cesare.pautasso}@usi.ch

Abstract. Producing initial process models currently requires gathering knowledge from multiple process participants and using modeling tools to produce a visual representation. With traditional tools this can require significant effort and thus delay the feedback cycle where the initial model is validated and refined based on participants' feedback. In this paper we propose a novel approach for process model sketching by applying existing process mining techniques to a sample process log obtained directly from the process participants. To that end, we specify a simple natural language-like domain-specific language to represent process traces or fragments of process traces. We also illustrate the architecture of a live modeling tool, the Sketch Miner, implementing the proposed approach. The tool produces a draft visual representation of the control flow which is updated in real-time as the traces are written down. The draft model generated by the tool can later be refined and completed by the business analysts using traditional tools.

Keywords: Draft process model · Process mining ·
Process requirements · Textual modelling DSL

1 Introduction

One of the identified challenges of business process management in a recent survey is the involvement of people with different skills and background [1], such as process participants with business domain knowledge, the business analysts with process modelling knowledge, and the software engineers with IT background. In the requirements gathering phase for the implementation of a new process in Process Aware Information Systems (PAIS), or when trying to model/improve existing processes, the people with detailed knowledge about the AS-IS or TO-BE business process are the participants in the process. Although they have deep knowledge about the activities that they perform [15], they might lack global knowledge about the end-to-end process [3]. Furthermore, abstracting from the process instances and thinking on the process model level is not always straightforward, especially for people with little or no modelling experience.

Drafting a model of the process with traditional modelling techniques, requires a dedicated person to step in the role of a business analyst, gather

© Springer Nature Switzerland AG 2019
T. Hildebrandt et al. (Eds.): BPM 2019, LNBIP 360, pp. 3–19, 2019.
https://doi.org/10.1007/978-3-030-26643-1_1

the process knowledge from different participants and use a graphical editor to manually create the initial draft process model. This requires time and significant cognitive effort by the business analyst, thus causing a delay in obtaining feedback on the draft model by the actual process participants. A comparative study by Damij [5] has shown that the process participant's role when using flowcharting modeling techniques, boils down to observation. While when using more natural language like techniques, such as activity tables, the process participants are actively involved to ensure that their activities are correctly captured in the model. On the same note, the study by Ottensooser et al. [18] has shown that people with no formal training in business process modeling, understand better textual notations describing business processes, such as written use-cases. However, their understanding of the described business process increases by reading a BPMN diagram after having read the written use-cases. To summarize, the existing state of the art process modelling techniques do not empower nor motivate process participants to become directly involved in the initial drafting of the process model. Their only possible involvement is during the interviews with the business analyst, who is then responsible for abstracting the concepts and creating the general picture, which might not align with the reality. However, a potential misalignment will only be discovered after the first draft of the model is finished by the business analyst and shown to the participants, thus creating delays in the feedback cycle.

With the work presented in this paper we would like to get the process participants to become actively involved in drafting the model, thus speeding up the feedback cycle. To that end, we propose a simple Domain Specific Language (DSL) that would allow process participants to describe their user stories in a predefined textual format, similar to task lists written in natural language. By transforming these lists into a process log we leverage on existing process mining algorithms to discover the process described by the participants. This Model Sketching by Mining approach allows the mining algorithm to use the unified knowledge of different participants to deduce the control flow branches and to infer the presence of loops, and thus output in real-time an initial draft of the business process in the visual language of choice. The role of the creator of the initial draft model is transferred from the business analysts to the mining algorithm, while the business analyst steps in later, in the refinement and finalization of the model, which can be done with the traditional modeling tools. This does not mean that the business analyst cannot use the DSL to "take notes" and sketch the model while interviewing process participants. That types of involvement allow for a quick initial draft of the process model which can be used to facilitate the discussion and the validation of the model with the stakeholders. While traditionally process mining has been used for discovery of existing processes based on system logs [22], we propose to extend the use of existing mining algorithms to sketch processes out of possible user scenarios or hypothetical participant stories.

This paper is structured as follows. In Sect. 2 we discuss briefly the existing textual notations for modelling business processes and motivate the need of defining our own DSL for capturing process knowledge, which we present in Sect. 3. In Sect. 4 we describe the architecture of the Sketch Miner, the proof of concept tool for Model Sketching by Mining, which translates the DSL into complete process traces to be used as input to a mining algorithm. To show the use of the DSL and the tool, in Sect. 5 we model a travel reimbursement process, which we refer to when discussing the benefits and the limitations of our approach in Sect. 6. In Sect. 7 we discuss the related work, while in Sect. 8 we conclude the paper and present work that we plan to conduct in the future.

2   Textual Notations for Process Modeling

Using textual notations to generate visual models has been gaining on importance in the modelling languages that target primarily developers. For instance, in Ballerina[1], a programming language aimed at the implementation of microservices and API integration, the textual and the visual syntax are kept synchronized as they are being edited independently. There are also tools, such as PlantUML, which support textual modeling of many of the Unified Modeling Language (UML) visual diagrams which are aimed at documenting software systems artifacts. One of these UML diagrams, the Activity diagram, is intended as a graphical representation of workflows and as such can be used to model business processes. It can be modelled graphically with tools such as StarUML[2], but it can also be modelled textually with tools such as PlantUML[3], following the syntax rules for distinguishing between different diagram constructs. The syntax uses: (1) *keywords*, such as "start" and "stop" to denote the beginning and end of a diagram, "if", "then", "else", "elseif" for conditionals, "repeat", "while" for loops, or (2) *punctuation signs*, such as ":" to denote an activity, "( )" to denote a gateway, "| |" to denote swimlanes etc. A survey of textual notations for UML is available in [19].

In the domain of process modelling, the Business Process Execution Language (BPEL) is a textual XML based language for specifying process behaviour with no standard graphical visual notation[4]. It uses nesting of constructs to represent the control flow, using keywords to distinguish among different control structures such as `<sequence>`, `<flow>`, `<if>` etc. The target audience of this language are developers. Nitzsche et al. [17] have proposed $BPEL^{light}$ to describe interactions only as message exchanges, regardless of the interface definition, thus separating the business logic from the technical protocols and messaging infrastructure. $BPEL^{light}$ extends BPEL with the `<conversation>` and `<interactionActivity>` elements which group the interaction activities and thus simplify the original BPEL language, facilitating the modelling of business

---

processes at a more abstract level. However, also $BPEL^{light}$ does not provide a visual rendering of the process model.

The Business Process Model and Notation (BPMN) is a standard process modelling language, designed starting from a visual syntax and later enhanced with an XML-based serialization. Recently a textual representation of BPMN has been proposed in plantBPMN [11]. Its target audience are developers, and potentially business analysts, and thus it supports a large number of BPMN constructs. It requires knowledge of BPMN terminology as specific keywords are used for all the node constructs such as "pool", "start", "start timer", "split", "task" etc. It also includes symbols that mimic the graphical constructs, such as the "->" for denoting the sequence flow.

The above mentioned DSLs require the users to learn and remember a syntax with different constructs, or rules for expressing different visualizations. As these rules are often similar to expressions used in programming languages, they might be intuitive for developers, but not necessarily for our primary target audience, i.e., process participants. The process participants would use the DSL for model sketching by mining only when mapping their everyday activities to a process, which is expected to happen rarely in their career. Thus, the textual DSL to be used by them should be as simple as possible and as close to natural language as possible, so that it can be explained and memorized fast, with no need of specialized training. The aim of such DSL is not to generate an executable process model, but rather to rapidly paint a model sketch for discussion purposes by inferring a process template from multiple scenarios representing individual instances. To do so, it does not need the full expressiveness of an executable process modeling language, such as BPMN or BPEL. Other authors have also pointed to the benefits of a simpler notation when gathering requirements [14]. By using mining to deduce the control flow constructs, we can avoid having to specify them in the textual DSL, and thus propose a simpler DSL than the ones mentioned above.

3   A DSL for Process Model Sketching by Mining

When the goal is obtaining an initial draft model fast, an important decision to make is what is the minimal necessary user input in order to maximize what can be expressed in the output. The desired output in our case is a draft process model, for which the required minimal input are the names of the activities in the process and the sequence in which they have to be completed. Although the DSL can be used by all involved parties, the primary targeted users are the process participants, who are generally not computer scientist, and thus are not familiar with the syntax of structured programming languages, and have limited knowledge of process modelling languages and process modelling. Therefore, one of the requirements for the DSL is that it should be as simple as possible and as close to natural language as possible. Additionally, as we propose using a mining algorithm to automatically generate the initial sketch of the process model, the input to the algorithm needs to mimic a set of traces of process instances. To

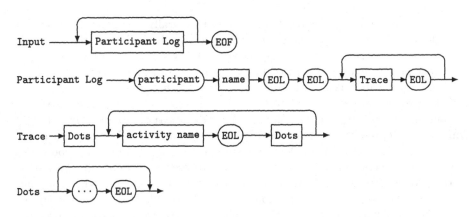

Fig. 1. Model sketching by mining DSL grammar definition (EOL indicates the end of the text line; EOF the end of the input file).

get the full model, a complete process instance trace is required for each possible path that can be taken in the process. There are several issues with obtaining such traces directly from the user. First of all, although it is highly likely that individual process participants are at least aware of the activities that they need to perform and the immediate predecessor/successor activities, they may lack knowledge of the end-to-end process, and thus all the possible paths in the process. Another issue is that stating all the possible paths becomes repetitive when there are many alternative paths which have many shared activities.

The standard IEEE format for event logs used in process mining is the eXtensible Event Stream (XES) format[5]. XES is a tag-based language, where the information about the events belonging to a determined process instance lie inside a trace element, which contains one to many event elements. Different attributes can be defined for the event element, the timestamp and the activity name being the most frequently used ones. Although XES is human readable, it requires the knowledge of XML syntax, the specific tags and their meaning and is rather verbose for human users to write manually. Thus, we could not use this standard as a format for the user input, but nonetheless we used it as a target into which our DSL can be transformed.

Namely, in the DSL for Model Sketching by Mining we require the user to write the name of each new activity in a new line, in the order in which the activities are completed. No time-stamp is required as the sequential order is deduced from the order in which the activities' names are written. While system event logs would have a process instance ID and a participant ID associated with each event, in our DSL a new empty line in the text separates traces of different process instances, and the **participant** keyword followed by the name of a participant denotes who performs the activities listed after naming the participant. In Fig. 1 we provide a formal definition of the syntax of the DSL specifying

---

[5] http://www.xes-standard.org.

the correct usage of the language constructs. Essentially, the allowed values in a single line of text are: (1) the `participant` keyword followed by the name of a participant; or (2) the name of the activity (empty spaces are allowed), or (3) an empty line (to mark the start of a new process instance); or (4) the dots "..." symbol (to indicate unknown fragments). While the names of the activities are sufficient to capture any complete trace of a process instance whose activities are performed by the same participant, we need the `participant` keyword as a construct to represent the hand-over of activities between participants. Furthermore, we need a construct to allow shortening the log by avoiding the repetition of the same sequences of activities in traces of process instances of different paths in the process. To that end, we introduce the "..." symbol. This symbol at the same time tackles the above mentioned issue of possible lack of knowledge of what the other participants are doing, which gives rise to the need to represent in the DSL fragments of process instance traces. The "..." symbol is essentially a placeholder for missing parts of the trace which are defined elsewhere. The precise semantics of the "..." symbol depends on its position relative to the start or end of the process instance trace, i.e., relative to the empty line. Following are the possible types of fragments of process instance traces and their semantics:

- Process instance trace that ***does not contain*** the "..." symbol: a sequence of activities which are all known to the user and which show a complete possible execution path of the process from the start to an end;
- Process instance trace that ***starts with*** the "..." symbol: the start of the process instance is not known to the user, or has been defined in the trace of other process instances. This fragment of a process instance trace states a sequence of activities that leads to the end of the process;
- Process instance trace that ***ends with*** the "..." symbol: the end of the process instance is not known to the user, or has been defined in the trace of other process instances. This fragment of a process instance trace states a sequence of activities that follows from the start of the process;
- Process instance trace that ***contains*** the "..." symbol: there is a part of the process instance that is not known to the user, or that has been defined in the trace of other process instances. This fragment of a process instance trace states a sequence of activities that starts and ends the process, but skips the middle of the process instance trace;
- Process instance trace that ***starts and ends with*** the "..." symbol: this type of trace is used as a fragment to fill in the placeholders in other traces in order to complete them. It does not correspond to a separate process instance.

4   Sketch Miner - A Tool for Model Sketching by Mining

As a proof of concept for the applicability of the proposed approach to use mining for sketching process models, we have designed the Sketch Miner, a tool which takes as input a process described with the DSL presented in Sect. 3 and provides as an output a draft BPMN model of the process (Fig. 2).

The tool first expands the traces written in the DSL to obtain the complete traces of all the possible paths that can be taken in the process. Namely, the DSL user input is parsed and each time the "..." symbol is encountered, depending on its position, the algorithm performs one of the following actions:

- If the "..." symbol is at the start of the process instance trace the algorithm searches for the first activity which is after the "..." symbol in all the other expanded process instance traces. When it finds it, it takes all the activities which precede it, thus creating one or more missing process fragments, which are then used to expand the initial process instance trace;
- If the "..." symbol is at the end of the process instance trace the algorithm searches for the last activity which is before the "..." symbol in all the other expanded process instance traces. When it finds it, it takes all the activities which succeed it, thus creating one or more missing process fragments, which are then used to expand the initial process instance trace;
- If the "..." symbol is in the middle of the process instance trace the algorithm searches for the first activity which is before the "..." symbol and the first activity which is after the "..." symbol in all the other expanded process instance traces. When it finds both these activities in the correct order, it takes all the activities which are between them, thus creating one or more missing process fragments, which are then used to expand the initial process instance trace.

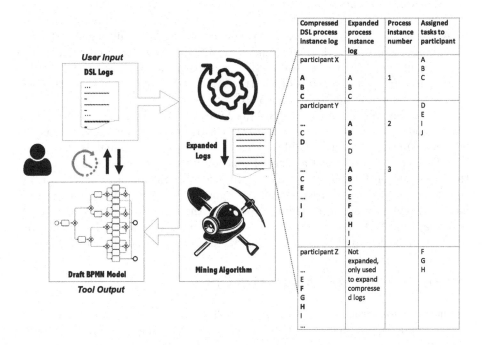

Fig. 2. Architecture of the Sketch Miner

As the trace expansion algorithm acts recursively and searches all expanded traces, if different sequences are identified as a match in different process instance traces, then they will all be used to expand the analyzed trace resulting with multiple expanded traces per one compressed trace. A simple example of how compressed traces are expanded by the Sketch Miner before passing them to the mining algorithm is provided in Fig. 2. As evident from the example, the assumption used by the expansion algorithm is the existence of at least one common activity between the compressed trace and the other traces, as the algorithm uses the common activity to identify the missing part of the path.

In order to enable automatic deduction of the activity role mapping, while keeping the DSL as simple as possible, we have introduced the constraint that the participant starting the process, when writing the traces, can only write down the activities (s)he is responsible for, using the "..." symbol when necessary. That way the algorithm can assign the activity to the first participant that mentions it. The other participants in the process, when using the "..." symbol, should state one activity preceding and/or following their activities depending on the position of the "..." symbol (beginning, end or middle of a process instance trace). In Fig. 2 there is a simple example showing which of the activities are assigned to which participant following the above mentioned constraints. Thus, we use the minimal assumption that process participants know all the activities they are responsible for as well as at least one activity preceding/following their activities so that log fragments of different participants can be connected. Nonetheless, participants may be aware of additional activities and they are free to state them.

The requirements for embedding a process mining algorithm within the Sketch Miner architecture involve the existence of an API to automatically feed the mining algorithm with the expanded traces. Additionally, users typing the traces should not have to wait to see the resulting model, but the model should be updated *live* as new entries of the traces are added. This could be achieved with an incremental mining approach [16].

The current version of the Sketch Miner uses the Alpha algorithm for mining the expended traces and produces as output a BPMN diagram, serialized as an SVG image using the dagre-d3 library[6], so that it can be immediately displayed in a Web browser. Aiming at a proof of concept of this novel modeling approach, the validation of different mining algorithms for its implementation was not in the scope of this work. Furthermore, the approach of using mining for process model sketching is not intended to depend on the target language and can be potentially applied to other process modelling languages, such as Petri Nets, or proprietary flowcharting languages. The Sketch Miner is available at Github[7].

5   Travel Reimbursement Use-Case

To show the use of the DSL proposed in Sect. 3 we will use the example of a travel costs reimbursement process. In this process, after returning from a business trip,

---

[6] https://github.com/dagrejs/dagre-d3.
[7] design.inf.usi.ch/sketch-miner.

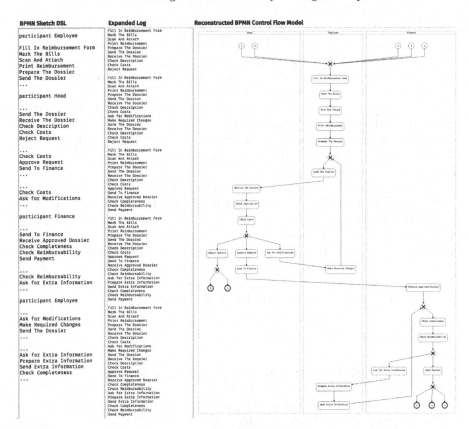

Fig. 3. Model sketching by mining in the travel reimbursement use-case

Employees need to fill in the reimbursement request form, mark the bills with the expense number, scan and attach them to the form, print the form, and together with the bills send it to their Head of Department for approval. After receiving the dossier the Head of Department checks the travel description and the stated costs, and can reject the request, or where needed ask the Employee to perform changes and resend the dossier. When the dossier is approved the Head of Department sends it to the Finance Department. After receiving it the Finance Department checks whether the submitted dossier is complete and whether the reported costs are reimbursable. If all the controls are passed successfully, the Finance Department sends the payment to the Employee. Otherwise they contact the Employee inquiring for additional information/documentation and re-performing the checks.

In the above described use-case there are three participants, the Employee, the Head of Department and the Finance Department. Probably none of them would know in detail what the other participants are doing as part of the described process, but we can assume that they all know at least the activity preceding/following the activities they are involved in. For instance, as can be seen in the left-hand side of Fig. 3, the Employees might not know the detailed steps that are taken by the Head of Department or the Finance Department, and thus they simply mark them with the "..." symbol. They know that some checks happen after they send the documents and that they can be asked to modify the request or to provide additional information, but they might not know the precise activities. On the other hand, the Head of Department can decide between three different paths after checking the documents, which is why (s)he specifies three different process instance traces. As two of the possible paths do not lead to an end of the process, they are placed in between the "..." symbols. On the other hand, the use of the "..." symbol at the beginning of the process instance traces allows the Head of Department to only state the new activities in the last two process instances, without repeating the activities already stated in the previous instance traces. Last but not least, the Finance Department, after performing the checks, can choose between asking for more information or sending the reimbursement payment, thus it only states two process instance traces. There are five possible paths assembled by the DSL log expansion, as evident in the central part of Fig. 3, and they are all passed to the mining algorithm in order to obtain the draft model presented in the right-hand side of Fig. 3.

6   Discussion

In real world scenarios, a written natural language description of the process is not always available, or when it is available it is not always up to date. In the travel reimbursement use-case mentioned in Sect. 5, with the traditional process modeling approach, such situation would require the business analyst to interview the three process participants in order to make the first draft model of the process. Then (s)he would need to identify the relevant activities at a meaningful level of granularity and name them. Bear in mind that in this toy example process there are already 20 activities, while in real world processes there can be many more. Then the business analyst, who is required to have prior knowledge of the visualization and the semantics of different BPMN constructs, or another process modelling language, should identify the divergence/convergence in the control flow and any possible loops. In the use-case model there are four exclusive gateways connected by two loops. Identifying them requires a cognitive effort, that as can be seen in Table 1, the business analyst is not spared of when using existing textual process modeling languages (described in Sect. 2). However, with our Model Sketching by Mining approach that part of the work is done by the mining algorithm in order to get the first draft of the process model. With this approach the active role of the business analyst can be postponed to after the first draft of the model, when (s)he would need to refine the draft model based on the discussion and the feedback from the process participants.

The Model Sketching by Mining approach enables the process participants to become directly involved in the modeling effort by writing the traces, regardless of the fact that they might lack the process modelling language knowledge. The DSL encourages them to think in terms of sequences of work units that they

Table 1. Textual process modelling languages comparison

	Target users	Users identify and name activities	Users identify the control flow	Modelling language support	Output format type	Model visualized
PlantUML	Developers	Yes	Yes	Full	PNG, SVG	On request
plantBPMN	Developers	Yes	Yes	High	BPMN	On save
BPELight	Business analysts	Yes	Yes	Full	BPEL	N/A
Sketch Miner	Process participants	Yes	No	Basic	SVG	Real time

perform, and how they should compose them into activities and name those activities in a manner that is meaningful to them. For instance, in the travel reimbursement process described in our use-case, for a business analyst the "Scan and attach the bills" step might not seem important as (s)he might not be aware of the average number of bills per request. Thus, (s)he might compose it together with the previous step "Mark the bills with the expense number" into a single activity. However, for the employees it might be important to single out this step as a separate activity as it is time consuming and they might want to keep track of the time they spend scanning. Providing meaningful names to the activities is also not an easy task for a business analyst, but it might come more natural for the actual participants executing the activities.

6.1 Usability vs. Expressiveness Trade-Off

The trade-off between usability and expressiveness is not inherent only to the domain of business process modelling [10]. The fundamental reason for this trade-off is the fact that greater expressiveness requires more language constructs, which hinders the usability as the users need more time and effort to learn the language. In existing textual process modeling languages the user typically needs to write text which mimics the graphical constructs of the modelling language (e.g., "->" to denote an edge in plantBPMN) or use the terminology of the modelling language (e.g., pool, task etc. in plantBPMN, fork in PlantUML) in order to obtain the visual model. This requires the user to have prior knowledge of the visual process modelling language. Usability, and fast learnability as part of it, is particularly important in our approach as we are primarily targeting process participants who are not frequently exposed to the visual process modelling language in their daily job. Thus, we do not aim at providing full support for all BPMN constructs. BPMN is a very expressive and rather complex language with over 100 constructs. As such providing a full support in our DSL would be likely to increase the complexity of the syntax, and drift the DSL design away from the original concept of using mining to infer the structure of the process as much as possible, as opposed to using the DSL to give a textual representation of the same, as in the case of PlantUML or plantBPMN. This is an important trade-off, which we have resolved favouring simplicity to enable process participants who have no knowledge of process modeling to state their user stories [14]. Furthermore, in order to both facilitate the learning of the basic BPMN constructs, and to provide fast feedback to the users, we have opted for generating the visual output at real time, as the users are typing their stories using the DSL. As evident in Table 1, this is not the case with existing textual process modeling languages. Namely, in PlantUML the textual model gets synchronized with the visual model only upon request, while plantBPMN generates a file in a BPMN XML format which then needs to be imported into a dedicated tool, such as Signavio[8] or an Eclipse plug-in, for visualization and further editing.

Even though we favour the DSL simplicity to expressiveness, we still aim to empower the process participants to tell their story by writing activity traces from their perspective. To that end, we have introduced the "..." symbol as a placeholder for parts of the process performed by others, and thus unknown to whoever is writing the story. The use of the "..." symbol is also meant to make the writing down of traces more time efficient, as it allows the users to avoid repeating sequences of activities which they have already written down. This is especially handy when there are alternative flows. For instance, in our travel reimbursement use-case, the Head of Department needs to make a decision whether to reject the request, ask for modification or send it to the Finance Department. Before making such decision the Head of Department would need to receive the reimbursement request dossier, check the travel description and check the stated costs. Writing down this sequence of activities three times would be too repetitive, so the Head of Department, when using the DSL, can avoid doing so by simply stating the above mentioned sequence in the first process instance trace and then starting a new instance trace using the "..." symbol followed by the "check the state costs" activity (see Fig. 3).

6.2 Potential Improvement of Modeling Efficiency

Bandara et al. [2] have identified the *modelling methodology* and the *modelling tool* as two factors impacting the success of a process modelling project. While the methodology refers to the modelling approach being followed (e.g., how is the requirements and information gathering phase performed), the modelling tool refers to the software being used for the design of the models. Each of these factors has its related costs in terms of efficiency and cognitive load. When modeling business processes, we can differentiate between the cognitive load for (1) identifying and naming the process activities, and (2) identifying the correct topology of the control flow graph. The intrinsic cognitive load of people depends on their prior knowledge and the complexity of the task that they need to perform [21]. It has been shown that, when it comes to understanding visual models, readers first identify smaller submodels and later connect everything together [13]. Our DSL design takes advantage of this by allowing participants

---

[8] https://www.signavio.com.

to specify sub-models they have first-hand knowledge about. These are then assembled by the mining algorithm to build the end-to-end process model. We expect this approach should decrease the cognitive load of the business analysts thanks to the use of a mining algorithm for reconstructing the control flow graph, which is an especially complex task in larger process models. There are various techniques for measuring the cognitive load, such as self-reported scales about the mental effort, or the difficulty of the tasks, as well as through response time [6], or eye movement tracking and pupillary response [4]. We plan to conduct such experiments in the future.

The type of *modelling tool* on the other hand, in addition to the cognitive load, can also impact the *time efficiency* of the modelling task per se. In a graphical editor a modeler would need to select the correct constructs from the available palette, place them in the modeling space, frequently using the drag&drop functionality, connect them in the correct order, and type the name of the activities. In the Sketch Miner, the user still needs to type the name of the activities as when working with the graphical editor. However, there is no drag&drop involved, thus there is no repositioning of the cursor within the activity shapes as all names are written on different lines of the same text editor. In other words, for equally experienced users we expect our textual entry to be more efficient than drawing graphical models, as argued in [11,12]. In the future, we plan to conduct controlled experiments where one group is asked to use the Sketch Miner and another group is asked to use a graphical editor to construct the same model with only the core BPMN constructs. By limiting the use of the BPMN constructs in the graphical editor we ensure that any potential overhead is not caused by the complexity of BPMN as a modeling language.

6.3 Limitations

One limitation of our current tool is that it uses the assumption that the participants know at least the last activity preceding/following their involvement in the process. This assumption requires that different participants use the same name to identify such activities. However, one of the main disadvantages of the natural language is its ambiguity. To deal with this limitation in future versions of the Sketch Miner we can leverage on work done in the creation of domain ontology and semantic annotation of process models expressed in BPMN. For instance, in [7] the authors propose using natural language parsing combined with information content similarity for generating suggestions for the semantic annotation of business process elements. We can use similar approach for suggestions or auto-completion of activity names. Another complementary solution is to enable collaborative editing of the traces so that participants can input them at the same time and resolve conflicts as soon as they appear.

As supporting parallel flows is only planned for future extensions of the Sketch Miner, we currently do not introduce the explicit notion of an instance id, which could be an approach for dealing with situations when work done by two different process participants is done in parallel.

Our current approach can lead to over-fitting, i.e., the automatically derived model might allow for process instances which are not described in the DSL. To deal with this limitation in the future we plan to generate the traces of such over-fitting instances and present them to the user so that the process participants can decide if some of those traces should be excluded from the model.

7   Related Work

Visual vs. textual modelling of processes has been long studied. Damij [5] studies the appropriateness of flowcharts vs activity tables for capturing the reality of a process using two case-study processes. Her work advises business analysts to start with the activity table and then transform the table into a flowchart. Otten-sooser et al. [18] use an experimental study with different types of participants to evaluate the impact of modeling with BPMN vs modelling with written use-cases on the understandability of the process. As Damij, Ottensooser et al. also advise to start with the textual technique. Namely, they show that the process under-standing of all participants benefited from reading the textual model, while only BPMN trained participants benefited from the visual model. However, they also observed that untrained participants who read the BPMN model after reading the textual model did improve their understanding based on the BPMN model. These studies have inspired our approach of using a textual DSL to automati-cally generate a visual BPMN model. Effektif[9] started with an idea similar to ours, i.e., hiding complexity from the users by allowing them to create task lists by simply naming tasks while the tool would create the corresponding visual BPMN task constructs. Users would then use Signavio as a standard graphical editor to modify the control flow topology, e.g., by connecting the tasks and introducing the appropriate gateways. Their goal was to facilitate the automa-tion of simple processes so that it can be done by any process participant, even without technical background. In our work we aim at simplifying and speeding up the creation of the initial model sketch by also deducing and drawing the control flow constructs, and not only the task constructs as in the case of Effek-tif. However, as in Effektiv the draft model generated with our approach may also need to be refined using a traditional graphical editor.

When it comes to related work in using process traces, in Test Driven Mod-elling (TDM) [23] for declarative process models, traces are used for creating test cases. A test case is a complete trace of a process instance that takes the form of a list of a sequence of activities that has to be supported by the defined process model. The completeness of traces requirement is relaxed in the recent work in [20]. However, in TDM traces are used for validation of an existing process model, while our primary goal is the sketching of the process model itself. In the past, scenarios and process fragments, called example runs [3] or oclets [9], have been used to create process models. While we propose a DSL for stating such process fragments, in [3] labelled partial orders and in [9] Petri Nets oclets are used. For the composition of the fragments we use a mining algorithm,

---

[9] https://www.signavio.com/post/introducing-effektif/.

while in the mentioned works the domain experts need to do the composition using composition operators (sequence, alternative, iteration, concurrency) [3] or combining actions enabled for extension [9].

Instead of the traditional process mining approach, whose output is automatically discovered process model, recent approaches tend to include the domain experts in the discovery of the process models, by allowing the user full control over the creation of the process model and simply providing suggestions regarding next activities based on the probabilities discovered with process mining algorithms [8]. While our work also requires domain experts' input, it does not require the existence of real-world, system generated process execution logs, thus it can also be applied for processes which are still not supported by PAIS.

8   Conclusions and Future Work

Gathering requirements for the design of a new PAIS can be a time consuming task as it requires the cooperation of software technology experts with business domain experts. In a traditional approach, the requirements gathering phase starts with interviews with process participants conducted by a business analyst who, based on those interviews, needs to sketch a process model to be discussed and agreed upon with the process participants. To facilitate the work of the business analyst, in this paper we have proposed an approach for process model sketching by mining activity log written down directly by the process participants using a textual DSL we have designed for this purpose. The DSL lists traces with activity names on separate lines and it uses only one keyword to specify participant names and one symbol to indicate trace fragments. We have also presented the Sketch Miner, a proof of concept implementation of the approach, using BPMN for the model visualization. The novelty of this approach is that the control flow is deduced automatically by a mining algorithm, based on the participants' user stories, and the draft process model is rendered in real-time as the users type in the activity traces. In a traditional approach the discovery of the control flow is done mentally by the business analyst who then needs to use graphical tools to represent it.

While working on the DSL in the future we plan to investigate how far we can go with increasing the expressiveness of the language, without making it too complex to be learned and effectively used by process participants, while utilizing as much as possible the mining technique to deduce the topology of the control flow graph. We also plan to empirically validate the approach. Namely, we expect that this new model sketching approach can speed up the feedback cycles in the process design as a result of the automatic process model generation which could reduce the cognitive load of the business analyst. However, we will need to run controlled experiments to systematically validate and quantify these expected benefits.

References

1. Alotaibi, Y., Liu, F.: Survey of business process management: challenges and solutions. Enterp. Inf. Syst. **11**(8), 1119–1153 (2017)
2. Bandara, W., Gable, G.G., Rosemann, M.: Factors and measures of business process modelling: model building through a multiple case study. Eur. J. Inf. Syst. **14**(4), 347–360 (2005)
3. Bergenthum, R., Desel, J., Mauser, S., Lorenz, R.: Construction of process models from example runs. In: Jensen, K., van der Aalst, W.M.P. (eds.) Transactions on Petri Nets and Other Models of Concurrency II. LNCS, vol. 5460, pp. 243–259. Springer, Heidelberg (2009). https://doi.org/10.1007/978-3-642-00899-3_14
4. Buettner, R.: Analyzing mental workload states on the basis of the pupillary hippus. NeuroIS **14**, 52 (2014)
5. Damij, N.: Business process modelling using diagrammatic and tabular techniques. Bus. Process Manag. J. **13**(1), 70–90 (2007)
6. DeLeeuw, K.E., Mayer, R.E.: A comparison of three measures of cognitive load: evidence for separable measures of intrinsic, extraneous, and germane load. J. Educ. Psychol. **100**(1), 223 (2008)
7. Di Francescomarino, C., Tonella, P.: Supporting ontology-based semantic annotation of business processes with automated suggestions. In: Halpin, T., Krogstie, J., Nurcan, S., Proper, E., Schmidt, R., Soffer, P., Ukor, R. (eds.) BPMDS/EMMSAD-2009. LNBIP, vol. 29, pp. 211–223. Springer, Heidelberg (2009). https://doi.org/10.1007/978-3-642-01862-6_18
8. Dixit, P.M., Verbeek, H.M.W., Buijs, J.C.A.M., van der Aalst, W.M.P.: Interactive data-driven process model construction. In: Trujillo, J.C., et al. (eds.) ER 2018. LNCS, vol. 11157, pp. 251–265. Springer, Cham (2018). https://doi.org/10.1007/978-3-030-00847-5_19
9. Fahland, D.: Oclets – scenario-based modeling with petri nets. In: Franceschinis, G., Wolf, K. (eds.) PETRI NETS 2009. LNCS, vol. 5606, pp. 223–242. Springer, Heidelberg (2009). https://doi.org/10.1007/978-3-642-02424-5_14
10. Freitas, A., et al.: Querying heterogeneous datasets on the linked data web: challenges, approaches, and trends. IEEE Internet Comput. **16**(1), 24–33 (2012)
11. Freund, N.: Development of a text-based representation of BPMN models. Master's thesis, Leibniz Universität Hannover, Hannover, Germany (2018)
12. Grönninger, H., Krahn, H., et al.: Textbased modeling. In: Proceedings of the 4th International Workshop on Software Language Engineering (2007)
13. Gruhn, V., Laue, R.: Reducing the cognitive complexity of business process models. In: 2009 8th IEEE International Conference on Cognitive Informatics, pp. 339–345. IEEE (2009)
14. Havey, M.: Keeping BPM simple for business users: power users beware. BPTrends, January 2006
15. Luftman, J.: Assessing it/business alignment. Inf. Syst. Manag. **20**(4), 9–15 (2003)
16. Masseglia, F., Poncelet, P., et al.: Incremental mining of sequential patterns in large databases. Data Knowl. Eng. **46**(1), 97–121 (2003)
17. Nitzsche, J., van Lessen, T., Karastoyanova, D., Leymann, F.: BPEL^{light}. In: Alonso, G., Dadam, P., Rosemann, M. (eds.) BPM 2007. LNCS, vol. 4714, pp. 214–229. Springer, Heidelberg (2007). https://doi.org/10.1007/978-3-540-75183-0_16
18. Ottensooser, A., Fekete, A., et al.: Making sense of business process descriptions: an experimental comparison of graphical and textual notations. J. Syst. Softw. **85**(3), 596–606 (2012)

19. Seifermann, S., Groenda, H.: Survey on the applicability of textual notations for the unified modeling language. In: Hammoudi, S., Pires, L.F., Selic, B., Desfray, P. (eds.) MODELSWARD 2016. CCIS, vol. 692, pp. 3–24. Springer, Cham (2017). https://doi.org/10.1007/978-3-319-66302-9_1

20. Slaats, T., Debois, S., Hildebrandt, T.: Open to change: a theory for iterative test-driven modelling. In: Weske, M., Montali, M., Weber, I., vom Brocke, J. (eds.) BPM 2018. LNCS, vol. 11080, pp. 31–47. Springer, Cham (2018). https://doi.org/10.1007/978-3-319-98648-7_3

21. Sweller, J.: Element interactivity and intrinsic, extraneous, and germane cognitive load. Educ. Psychol. Rev. 22(2), 123–138 (2010)

22. van der Aalst, W.: Process Mining: Discovery, Conformance and Enhancement of Business Processes. Springer, Heidelberg (2011). https://doi.org/10.1007/978-3-642-19345-3

23. Zugal, S., Pinggera, J., Weber, B.: Creating declarative process models using test driven modeling suite. In: Nurcan, S. (ed.) CAiSE Forum 2011. LNBIP, vol. 107, pp. 16–32. Springer, Heidelberg (2012). https://doi.org/10.1007/978-3-642-29749-6_2

Quasi-Inconsistency in Declarative Process Models

Carl Corea[✉] and Patrick Delfmann

Institute for Information System Research,
University of Koblenz-Landau, Koblenz, Germany
{ccorea,delfmann}@uni-koblenz.de

Abstract. The field of declarative process discovery comprises techniques for mining declarative constraint sets from event logs. While current techniques verify the relation of individual constraints to the log, they do not consider the interrelation between constraints. This can lead to logical contradictions between the discovered constraints. In this work, we introduce a new form of such contradictions entitled implicit inhibitors. In short, these are sets of constraints which will always be activated together, but demand contradicting reactions. In turn, such constraint sets can be denoted as quasi-inconsistent, as the contained constraints are unsatisfiable should they be activated together. We introduce a structured approach to detect and analyze quasi-inconsistencies in declarative process models and evaluate our approach through formal analysis and run-time experiments on real-life data-sets.

Keywords: Declarative constraints · Implicit inhibition · Declare

1 Introduction

Declarative process models consist of constraints which specify the behavior which company processes should adhere to. Process execution in declarative process models is thus all allowed behaviour within the set of constraints. The semantics of declarative constraints is mostly formalized with temporal logic, e.g. with modelling languages such as DECLARE [6,14]. For example, the DECLARE constraint $\text{CHAINRESPONSE}(a, b)$ imposes that if a task a occurs, it must be directly followed by a task b. Likewise, $\text{RESPONSE}(a, b)$ states, that if a task a occurs, it must be eventually followed by a task b. When utilizing declarative models, companies face numerous current challenges: In the scope of process discovery, current discovery techniques can yield sets of constraints which are unusable or confusing to modelers [7]. Also, human modelling errors or merging models in the scope of company mergers can yield erroneous models [3].

As an example, consider the constraint sets $\mathcal{C}_1$ and $\mathcal{C}_2$, defined via

$$\mathcal{C}_1 = \{\text{CHAINRESPONSE}(a, b) \qquad \mathcal{C}_2 = \{\text{CHAINRESPONSE}(a, b)$$
$$\text{NOTRESPONSE}(a, b)\} \qquad \qquad \text{CHAINRESPONSE}(b, c)$$
$$\text{NOTRESPONSE}(a, c)\}.$$

© Springer Nature Switzerland AG 2019
T. Hildebrandt et al. (Eds.): BPM 2019, LNBIP 360, pp. 20–35, 2019.
https://doi.org/10.1007/978-3-030-26643-1_2

In both cases, the task a is inhibited by (multiple) constraints which define which tasks must or must not follow. However, these constraints demand logically contradicting reactions to the occurence of task a. In turn, should the task a occur, the declarative process model cannot further be executed.

Motivation: Can't this be already solved with finite state automata? The observant reader might ask whether the above examples are not *inconsistencies* as defined in Di Ciccio et al. [7], and thus could already be detected by existing approaches such as automata products. In short, the above examples are not inconsistencies, but rather *quasi*-inconsistencies, explained as follows.

Di Ciccio et al. [7] have discussed the problem of inconsistent constraints that can be returned during process discovery. Those authors define inconsistency as a declarative process model which does not accept any execution trace, i.e. it is unsatisfiable. An example would be the constraint set C_3, defined via

$$C_3 = \{\text{PARTICIPATION}(a)$$
$$\text{CHAINRESPONSE}(a, b)$$
$$\text{NOTRESPONSE}(a, b)\}.$$

As can be seen, C_3 contains the constraint PARTICIPATION(a), which states that the task a <u>must</u> occur in every execution trace. In result, the constraints CHAINRESPONSE(a, b) and NOTRESPONSE(a, b) must also always be activated in any execution trace. However, this constellation is inconsistent, i.e. there cannot exist a trace that satisfies the model in C_3.

On the contrary, a model containing the constraints in C_1 or C_2 can accept an arbitrary amount of execution traces, namely any trace which does not contain the task a. For example, a trace *"bcbdbde"* would satisfy respective models in C_1 or C_2. Thus, these constraint sets are not inconsistent as defined in [7], but rather *quasi-inconsistent*. That is, certain tasks are implicitly inhibited by a set of contradictory constraints. Due to this different conceptualization, this implicit inhibition can however not be detected via automata products as in [7], as there can be a non-empty set of accepted traces. In result, this paper discusses a new form of problem in declarative process discovery. Intuitively, declarative process models should not contain sets of constraints as in C_1 or C_2, as the can potentially make models unusable. Yet, current discovery techniques can return such quasi-inconsistent constraint sets. Furthermore, as motivated above, such quasi-inconsistencies can currently not be detected by existing means such as automata products. This is underlined by our experiment results (cf. Sect. 5), where we analyzed real-life models and found more than 25.000 of such contradictory constraint sets as in $C1$ or C_2, which cannot be detected with the approach by Di Ciccio et al. [7]. In this work, we therefore introduce the notion of quasi-inconsistency and propose a first structured approach to detect and analyze all minimal implicit inhibition sets in declarative process models.

The remainder of this paper is as follows. Section 2 provides background information on declarative process models. In Sect. 3, we introduce the novel concept of *quasi-inconsistent subsets* and show how results from the scientific field of inconsistency measurement can be adapted to detect and analyze such subsets. In Sect. 4, we present an algorithm for the feasible computation of quasi-inconsistent subsets. The proposed capabilities for detection and analysis are evaluated in Sect. 5, followed by a conclusion in Sect. 6.

2 Background

Traditional process models define a clear imperative structure of how exactly company activities should be executed. To allow for more flexibility, declarative process models have received increasing attention [1,6,7]. Here, a declarative process model defines constraints which must be upheld or not be violated, and thus process execution is flexible within this set of constraints.

Definition 1 (Declarative Process Model). *A declarative process model is a tuple* $\boldsymbol{M} = (\boldsymbol{A}, \boldsymbol{T}, \boldsymbol{C})$, *where* $\boldsymbol{A}$ *is a set of tasks,* $\boldsymbol{T}$ *is a set of constraint templates, and* $\boldsymbol{C}$ *is the set of actual constraints, which instantiate the template elements in* $\boldsymbol{T}$ *with tasks in* $\boldsymbol{A}$.

In this paper, we consider DECLARE [14], which is a widely acknowledged declarative process modelling language and notation. DECLARE allows to define constraints by using predefined templates and passing tasks as parameters to respective templates, cf. the examples in Sect. 1. In this way, modelers can use the rather intuitive templates to define constraints, with the formal semantics "hidden" from the user. Formally, the semantics of DECLARE can be defined with temporal logic [1,4]. This allows to use the amenities of temporal logic checking, as well as to create custom DECLARE constraint templates.

We define the semantics of DECLARE constraints with LTL_p [13], a linear-time temporal logic with past. An LTL_p formula is given by the grammar

$$\varphi ::= a | (\neg \varphi) | (\varphi_1 \wedge \varphi_2) | (\bigcirc \varphi) | (\varphi_1 \mathbf{U} \varphi_2) | (\ominus \varphi) | (\varphi_1 \mathbf{S} \varphi_2).$$

Each formula is built from atomic propositions $\in \mathbf{A}$ (relative to a declarative process model), and is closed under the boolean connectives, the unary temporal operators $\bigcirc$ (next) and $\ominus$ (previous), and the binary temporal operators $\mathbf{U}$ (until) and $\mathbf{S}$ (since). Given a declarative process model $\mathbf{M} = (\mathbf{A}, \mathbf{T}, \mathbf{C})$, a sequence t (with length n) of tasks in $\mathbf{A}$, where $t(i)$ denotes the ith element of the sequence t, the semantics of LTL_p formulae are defined as follows:

$t,i \models \textit{True} / t,i \not\models \text{False}$ $\quad t,i \models a$ iff $t(i) = a$

$t,i \models \neg\varphi$ iff $t,i \not\models \varphi$ $\quad t,i \models \varphi_1 \wedge \varphi_2$ iff $t,i \models \varphi_1$ and $t,i \models \varphi_2$

$t,i \models \bigcirc\varphi$ iff $i < n$ and $t,i+1 \models \varphi$ $\quad t,i \models \ominus\varphi$ iff $i > 1$ and $t,i-1 \models \varphi$

$t,i \models \varphi_1 \mathbf{U} \varphi_2$ iff $t,j \models \varphi_2$ with $i \leq j \leq n$, and $t,k \models \varphi_1$ for all k s.t. $i \leq k < j$

$t,i \models \varphi_1 \mathbf{S} \varphi_2$ iff $t,j \models \varphi_2$ with $1 \leq j \leq i$, and $t,k \models \varphi_1$ for all k s.t. $j < k \leq i$

From the above syntax and semantics, we furthermore derive $\varphi_1 \vee \varphi_2$ as $\neg(\neg\varphi_1 \wedge \neg\varphi_2)$, $\varphi_1 \rightarrow \varphi_2$ as $\neg\varphi_1 \vee \varphi_2$, $\Diamond\varphi$ as $True\mathbf{U}\varphi$ (which indicates that φ will eventually hold true, possibly later and not directly following $t(i)$), $\Diamond\!\!\!-\varphi$ as $True \mathbf{S} \varphi$ (which indicates that φ holds true sometime before $t(i)$, but not necessarily directly before $t(i)$), and $\Box\varphi$ as $\neg\Diamond\neg\varphi$ (which indicates that there is no future $t(i)$ which does not satisfy φ).

Based on such LTL_p formulae, the semantics of individual DECLARE constraints can be defined. For instance, the exemplary constraints used in $\mathcal{C}_1$ and $\mathcal{C}_2$ are defined as CHAINRESPONSE$(a, b) \equiv \Box(a \rightarrow \bigcirc b)$, NOTCHAINRE-SPONSE$(a, b) \equiv \Box(a \rightarrow \neg \bigcirc b)$, NOTRESPONSE$(a, b) \equiv \Box(a \rightarrow \neg\Diamond b)$. A standard set of Declare templates and corresponding semantics have been defined derived from the work of [8]. Please see [1,7] or [12] for further details.

An interesting gist about such constrains is that DECLARE seems to capture *activation-response* relations between tasks. For instance, CHAINRESPONSE(a, b) can be interpreted such that, *if* there is an activation a, then this entails a reaction $\bigcirc b$. Therefore, following [2], we use the notion of reactive constraints, which make the activation and reaction semantics of LTL_p constraints explicit.

Definition 2 (Reactive Constraints [2]). *Given a declarative process model* $\boldsymbol{M} = (\boldsymbol{A}, \boldsymbol{T}, \boldsymbol{C})$, *and a constraint* $\in \boldsymbol{C}$ *with activation* α *and reaction* φ, *a reactive constraint (RCon)* Ψ *is a pair* (α, φ). *We denote* $\Psi = (\alpha, \varphi)$ *as* $\alpha \Rightarrow \varphi$. *We say that* α *activates the constraint and the reaction* φ.

Table 1 provides an overview of DECLARE constraints used in this work, as well as the corresponding RCon and activation. Please refer to [2,7] for a further discussion and classification of activations in DECLARE constraints.

Table 1. Reactive constraints corresponding to exemplary DECLARE constraints

Constraint	Reactive constraint	Activation
RESPONSE(a, b)	$a \Rightarrow \Diamond b$	a
CHAINRESPONSE(a, b)	$a \Rightarrow \bigcirc b$	a
ALTERNATERESPONSE(a, b)	$a \Rightarrow \bigcirc(\neg a \mathbf{U} b)$	a
PRECEDENCE(a, b)	$b \Rightarrow \Diamond\!\!\!- a$	b
CHAINPRECEDENCE(a, b)	$b \Rightarrow \ominus a$	b
ALTERNATEPRECEDENCE(a, b)	$b \Rightarrow \ominus(\neg b \mathbf{S} a)$	b
NOTRESPONSE(a, b)	$a \Rightarrow \neg\Diamond b$	a
NOTCHAINRESPONSE(a, b)	$a \Rightarrow \neg \bigcirc b$	a
NOTPRECEDENCE(a, b)	$b \Rightarrow \neg\Diamond\!\!\!- a$	b
NOTCHAINPRECEDENCE(a, b)	$b \Rightarrow \neg\ominus a$	b

In result, a quasi-inconsistency is present if we have a constraint set containing multiple RCons with the same activation, but contradictory reactions. In the following, we will show how such quasi-inconsistencies in declarative process models can be detected and analyzed.

3 Detecting and Assessing Quasi-Inconsistencies

3.1 Detection

As declarative constraints are inherently of reactive nature, they underly the principle of ex falso quodlibet: no conclusions can be made without knowledge of activation. As motivated in Sect. 1, this means that the exemplary constraint sets C_1 and C_2 are not inconsistent per se, as it is not known whether these constraints will actually be activated (i.e., there is no constraint like PARTICIPATION(a) which dictates the occurrence of a task a in an execution). In turn, there can be an arbitrary amount of traces that satisfy models as in C_1 and C_2, thus it is not possible to detect quasi-inconsistency by the existing means of automata products as in [7], which detects inconsistency as an empty set of acceptable input traces.

Thus, we present a novel means for detecting quasi-inconsistency. In the following, we use the RCon representation, but sometimes provide specific DECLARE templates for readability. Furthermore, let a constraint c, we denote $out(c)$ as the outcome of a constraint, i.e. φ of the respective RCon.

Definition 3 (Individual Constraint Activation). *A set of activations A activates an individual constraint $c : a \Rightarrow \varphi$ iff $a \in A$.*

Quasi-inconsistencies can arise, if we have a set of activations A', such that A' activates at least two different constraints, and these constraints have contradictory outcomes, e.g. in example C_1, the activation set $\{a\}$ activates two contradictory constraints. However, as the conclusions of some constraints might be an activation to other constraints themselves via transitive relations, the activation set A' might activate a multitude of constraints. In order to analyze quasi-inconsistencies, all these activated constraints must be considered.

Definition 4 (Constraint Set activation). *A set of activations A activates a set of constraints C iff $\forall c \in C : A \cup \{out(c)|c \in C\}$ activates c.*

Example 1. Consider the constraint set C_4, defined via

$$C_4 = \{a \Rightarrow b, b \Rightarrow c, c \Rightarrow d\}.$$

For each individual constraint, the activation set is simply the premise of the constraint, i.e. a is the activation set of the individual constraint $a \Rightarrow b$. Furthermore, the activation a also activates the entire set of constraints in C_4 via the transitive relations.

Given a declarative set of constraints, the introduced notions allow to define *quasi-inconsistent subsets.*

Definition 5 (Quasi-Inconsistent Subset). *For a constraint set C, the set of quasi-inconsistent subsets QI is defined as a set of pairs $(\boldsymbol{A}, \boldsymbol{C})$, s.t.*

1. $\boldsymbol{C} \subseteq C$

2. **A** activates **C**
3. $\boldsymbol{A} \cup \boldsymbol{C} \models \bot$

To clarify, we consider a set of activations **A**, which activate **C**. Then, the entirety of all activations and activated constraints is inconsistent. Our proposition of quasi-inconsistent subsets allows to determine the "inconsistent subsets" of arbitrary declarative constraints sets, by augmenting activations and thus determining those constraints which will (a) always be activated together, and (b) yield an inconsistency, should they be activated. Consequently, we define minimal quasi-inconsistent subsets analogously.

Definition 6 (Minimal Quasi-Inconsistent Subset). *For a constraint set C, the set of minimal quasi-inconsistent subsets MQI is defined as set of pairs t = (A, C), s.t.*

1. *t is a quasi-inconsistent subset in C*
2. *for any $t' \subset t$, where exactly one element is deleted from exactly one of the sets in t,: $t' \not\models \bot$*

A minimal quasi-inconsistent subset is a quasi-inconsistent subset which is minimal w.r.t. set inclusion, i.e., removing exactly one constraint resolves the quasi-inconsistency. As we are mostly interested in the distinct constraints which are quasi-inconsistent to each other, we use M^C to denote the set of constraints **C** from any $M \in$ MQI.

Example 2. Consider the following DECLARE constraint set $\mathcal{C}_5$, defined via

$\mathcal{C}_5 = \{$CHAINRESPONSE(a, b),   RESPONSE(b, d),   NOTCHAINPRECEDENCE(a, b)
          CHAINRESPONSE(d, e),   NOTRESPONSE(a, b),   CHAINRESPONSE(e, c)
          CHAINRESPONSE(b, c),   RESPONSE(a, b) NOTRESPONSE(a, c) $\}$

Then[1],

MQI$(\mathcal{C}_5) = \{M_1, M_2, M_3, M_4, M_5, M_6, M_7\}$

$M_1^C = \{$NOTCHAINPRECEDENCE(a, b), CHAINRESPONSE$(a, b)\}$

$M_2^C = \{$CHAINRESPONSE(a, b), NOTRESPONSE$(a, b)\}$

$M_3^C = \{$RESPONSE(a, b), NOTRESPONSE$(a, b)\}$

$M_4^C = \{$RESPONSE(a, b), NOTRESPONSE(a, c), CHAINRESPONSE$(b, c)\}$

$M_5^C = \{$CHAINRESPONSE(a, b), CHAINRESPONSE(b, c), NOTRESPONSE$(a, c)\}$

$M_6^C = \{$RESPONSE(a, b), RESPONSE(b, d), CHAINRESPONSE(d, e),
        CHAINRESPONSE(e, c), NOTRESPONSE$(a, c)\}$

$M_7^C = \{$CHAINRESPONSE(a, b), RESPONSE(b, d), CHAINRESPONSE(d, e),
        CHAINRESPONSE(e, c), NOTRESPONSE$(a, c)\}$

$M_1 = (\{a\}, \{$NOTCHAINPRECEDENCE(a, b), CHAINRESPONSE$(a, b)\})$

---

[1] M_2–M_7 are omitted due to space restrictions, but are analogously to M_1 (all with activation set $\{a\}$).

This example shows the minimal quasi-inconsistent subsets of the constraint set C_5. As can be seen, all such subsets implicitly inhibit certain tasks in an unsatisfiable way. Thus, in the example, should the task a occur, the resp. model is unsatisfiable and can thus not be used for simulation or to govern compliant process execution. Intuitively, declarative process models should therefore not contain such inhibiting subsets. Through our novel definition of quasi-inconsistent subsets, we are able to detect all such problematic subsets within a set of constraints. Furthermore, our definition of quasi-inconsistent subsets enables a further assessment of resp. subsets.

3.2 Analysis

In order to understand potential inconsistencies, companies should be provided with a careful analysis of detected quasi-inconsistencies. To this aim, results from the scientific field of inconsistency measurement can be adapted [10]. Inconsistency measurement is a discipline concerned with the analysis of inconsistent information. Here, the central object of study are quantitative measures, which allow to assign a numerical value to (elements of) a constraint set, with the informal meaning that a higher value reflects a higher degree of inconsistency. These measures can be distinguished into so-called *inconsistency measures*, and *culpability measures*. The former is used to assess the inconsistency of the entire constraint set, while the latter is used to assess the degree of blame that individual constraints carry in the context of the overall inconsistency. As some of these measures are based on set-theoretic principles, we propose to adopt these measures to analyze quasi-inconsistent subsets as follows.

Quasi-Inconsistency Measures. Let $\mathfrak{C}$ denote the universe of all declarative constraint sets. Then, an inconsistency measure $\mathcal{I}$ is a function

$$\mathcal{I} : \mathfrak{C} \to [0, \infty)$$

which assigns a non-negative real value to a constraint set, with the informal meaning that a higher value reflects a higher severity of inconsistency.

Following recent surveys [16,17], there are four measures based on minimal inconsistent subsets which have been proposed, namely the MI-*inconsistency measure*, the MI^C-*inconsistency measure*, the *problematic inconsistency measure* and the *mv-inconsistency measure*. Currently these measures are only defined for inconsistencies (and not for quasi-inconsistencies). To analyze the degree of quasi-inconsistency of declarative process models, these can easily be adapted to fit the use-case of quasi inconsistencies. For further evaluation, we present this for the example of the MI-*inconsistency measure*. We omit a detailled discussion of all measures due to space limitations.

Let C be a set of constraints and $\mathcal{A}(C)$ denote the tasks in a set C. Then, the adapted versions of the abovementioned measures are defined as follows.

Definition 7 (MQI-inconsistency measure). *Define the* MQI*-inconsistency measure via*

$$\mathcal{I}^Q_{\mathsf{MI}}(\mathcal{C}) = |MQI(\mathcal{C})|$$

This measure counts the number of minimal quasi-inconsistent subsets in $\mathcal{C}$.

Example 3. We revisit the constraint set $\mathcal{C}_5$ from Example 2. Then

$$\mathcal{I}^Q_{\mathsf{MI}}(\mathcal{C}_4) = 7$$

Culpability Measures. Next to assessing the degree of inconsistency for an entire constraint set, results from inconsistency measurement also allow to quantify the degree of inconsistency for individual constraints. This allows to pinpoint constraints with a high degree of blame for the overall inconsistency. Let $\mathfrak{c}$ denote the universe of all possible constraints, and $\mathfrak{C}$ the universe of declarative constraint sets. Then, a culpability measure Γ is a function

$$\Gamma : \mathfrak{C} \times \mathfrak{c} \to [0, \infty)$$

which assigns a non-negative number to a mapping of an individual constraint to a constraint set, and can thus assess the culpability that an individual constraint represents w.r.t. the constraint set. There are two culpability measures based on minimal inconsistent subsets which have been proposed, namely the *cardinality based culpability measure* $\Gamma_{\#}$, and the *normalized culpability measure* Γ_c [11]. Again, these can be easily adopted for the use-case at hand, which we show for the cardinality-based culpability measure.

Definition 8 (Cardinality-Based Culpability Measure). *Define the* cardinality based culpability measure $\Gamma^Q_{\#}$ *via*

$$\Gamma^Q_{\#}(\mathcal{C}, \alpha) = |M \in MQI(\mathcal{C})|\alpha \in M^C|$$

This measure counts the number of minimal quasi-inconsistent subsets that a constraint α appears in.

Example 4. We revisit the constraint set $\mathcal{C}_5$ from Example 2. Then

(i) $\Gamma^Q_{\#}(\mathcal{C}_5, \text{CHAINRESPONSE}(a, b)) = 4$ (vi) $\Gamma^Q_{\#}(\mathcal{C}_5, \text{RESPONSE}(b, d)) = 2$

(ii) $\Gamma^Q_{\#}(\mathcal{C}_5, \text{NOTCHAINPRECEDENCE}(a, b)) = 1$ (vii) $\Gamma^Q_{\#}(\mathcal{C}_5, \text{CHAINRESPONSE}(d, e)) = 2$

(iii) $\Gamma^Q_{\#}(\mathcal{C}_5, \text{NOTRESPONSE}(a, b)) = 2$ (viii) $\Gamma^Q_{\#}(\mathcal{C}_5, \text{CHAINRESPONSE}(e, c)) = 2$

(iv) $\Gamma^Q_{\#}(\mathcal{C}_5, \text{CHAINRESPONSE}(b, c)) = 2$ (ix) $\Gamma^Q_{\#}(\mathcal{C}_5, \text{RESPONSE}(a, b)) = 3$

(v) $\Gamma^Q_{\#}(\mathcal{C}_5, \text{NOTRESPONSE}(a, c)) = 4$

Culpability measures provide quantitative insight that can help companies to understand and resolve problems in their models [15]. The intuition here is that a higher culpability reflects a higher degree of blame that an individual constraint carries in the context of the overall inconsistency [9]. For example, the $\Gamma^Q_{\#}$ is essentially a scoring function which quantifies how many quasi-inconsistent subsets can be resolved, if a constraint is deleted.

4   Computation of Quasi-Inconsistent Subsets

The basis for the proposed detection and analysis are quasi-inconsistent subsets. In the following, we therefore propose an novel approach for the feasible computation of MQI. Algorithm 1 shows our approach to compute minimal quasi-inconsistent subsets for declarative constraints. As a central object of study, we utilize reactive constraints to construct a so-called *reactive entailment graph*.

4.1   Reactive Entailment Graph

Definition 9 (Reactive Entailment Graph). *Given a declarative process model $M = (A, T, C)$, its reactive entailment graph (REG) is defined as a graph $G = (A, E, \tau, n)$, where $A = A \cup \overline{A}$ are the tasks in M in two forms (with and without overline symbol), $E \subseteq A \times A$ is the set of directed edges between tasks in A, τ is a function $\tau : E \to T$ assigning an individual edge in E to a template type in T, and n is a function $n : E \to \mathbb{N}$ which assign a natural number to an edge to allow for multiple edges between the same tasks in A.*

The reactive entailment graph is a graph representation of reactive constraints. For example, given the declarative constraint CHAINRESPONSE(a, b), this can be represented as two nodes a and b, related by an edge of type ○. In the following, we ommit edge numbering for simplicity.

An important detail is that we include two "forms" of tasks, explained as follows. As can be seen in Table 1, one could argue there are essentially two types of declarative constraints. First, there are constraints such as CHAINRESPONSE, which are aimed to ensure that, should some event occur, then another event *must* occur (in a certain way). Then, there are other constraints such as NOTCHAINRESPONSE, which are aimed to ensure that, should some event occur, then another event *must not* occur (in a certain way). The reactive entailment graph captures these two types of *demanding* and *prohibiting* constraints, with the intuition that the overlined form of a task relates to a prohibition and vice versa. Then, the edges, respectively the edge types convey information on how exactly a task is demanded or prohibited, w.r.t a node which is the activation.

Example 5. We revisit the exemplary constraint set from Example 2. Then, this yields the following reactive entailment graph:

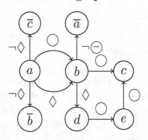

This graph encodes the relations between tasks of a declarative process model, as well as their relation type. For example, it can be seen that $a \Rightarrow \Diamond b$, and $a \Rightarrow \neg \Diamond \overline{b}$. This encodes that the activation a *demands* task b, resp. *prohibits* a later occurence of task b. An advantage of including two forms of tasks to encode the demanding, resp. prohibiting, nature of reactive constraints is an efficient way to scan the REG for potential inconsistencies, by searching for pairs of nodes n and n', where $n = \overline{n'}$, as will be discussed in the subsequent section.

The graph relations can be transformed back into the original constraints, where if $(\alpha, t_i) \in E$, then the original reactive constraint c_i is defined as $c_i = \alpha \Rightarrow \tau(\alpha, t_i)t_i$. For an edge $e \in E$, we denote the corresponding constraint as e^C. Given a path p being a sequence of edges in the REG, we denote the set of corresponding constraints captured by p as p^C.

4.2   Algorithm for Computing MQI in Declarative Constraint Sets

Following Definition 5, quasi-inconsistency can only occur if

1. There is at least one task Δ
2. Δ is the outcome of at least two constraints c_1 and c_2
3. $out(c_1) = \overline{out(c_2)}$

Furthermore, the constraints c_1 and c_2 have to be activated simultaneously, thus

4. c_1 and c_2 have the same activation set a

Algorithm 1 computes MQI of declarative constraint sets by exploiting the reactive entailment graph to search for subsets satisfying 1–4. In the following, we explain our algorithm based on the constraint set from Example 2 and the corresponding REG from Example 5.

Algorithm 1. Computation of minimal quasi-inconsistent subsets

    Input   : Set of constraints **C**
    Output: MQI(**C**)

1   $qmis \leftarrow \emptyset$;
2   $compConstraints = findComplements(C)$;
3   **foreach** n:$compConstraints$ **do**
4       $\alpha \leftarrow n.activation$;
5       $\omega \leftarrow n.reactionTask$;
6       $\mathbf{P} = findPaths(\alpha, \overline{\omega}) \cup findPaths(\overline{\omega}, \alpha)$;
7   **foreach** P:$\mathbf{P}$ **do**
8   **if** $\alpha \cup n \cup P^C \models \perp$ **then**
9             $mis \leftarrow mis \cup n \cup P^C$;

In line 1, a set to store minimal quasi-inconsistent subsets ($mqis$) is initialized. Then, we start by identifying all nodes n' of the REG which are a complement to another node n'' (line 2). In the example, there are three such cases, namely a vs $\bar{a}$, b vs $\bar{b}$ and c vs $\bar{c}$ (cf. the corresponding REG). Due to space limitations, we focus on $\bar{c}$ in the following, i.e., we assume the current iterated node $n_i = \bar{c}$. Its activation α is its predecessor in the REG, here a. The algorithm subsequently search for all shortest paths from α to the inverse of the current n_i via a breadth-first search, stored in P (i.e., in our example the algorithm searches for all shortest paths from a to c, and from c to a in the REG). We store possible paths from $\overline{n_i}$ to α, to cope with constraints such as precedence. Also, note that these can be transitive paths with multiple hops. As can be seen in the REG in Example 5, there are four paths from a to c. Subsequently, the algorithm verifies whether the constraints pertaining to a found path P contradict the original constraint $c_i = a \Rightarrow \neg\Diamond\bar{c}$. To this aim, we verify if $\alpha \cup c_i \cup P^C \models\perp$, in which case we have found a minimal quasi-inconsistent subset. In the example, the conditions verified in line 8 are respectively:

(1) $a \cup \text{NOTRESPONSE}(a, c) \cup \{\text{CHAINRESPONSE}(a, b), \text{CHAINRESPONSE}(b, c)\} \models\perp$

(2) $a \cup \text{NOTRESPONSE}(a, c) \cup \{\text{RESPONSE}(a, b), \text{CHAINRESPONSE}(b, c)\} \models\perp$

(3) $a \cup \text{NOTRESPONSE}(a, c) \cup \{\text{CHAINRESPONSE}(a, b), \text{RESPONSE}(b, d),$
       $\text{CHAINRESPONSE}(d, e), \text{CHAINRESPONSE}(e, c)\} \models\perp$

(4) $a \cup \text{NOTRESPONSE}(a, c) \cup \{\text{RESPONSE}(a, b), \text{RESPONSE}(b, d),$
       $\text{CHAINRESPONSE}(d, e), \text{CHAINRESPONSE}(e, c)\} \models\perp$

Note that the activation α is augmented in line 8 to allow for this detection of quasi-inconsistent subsets via Definition 5. Concluding the example, as all 4 cases return true, we have successfully found four *mqis* based on the reactive entailment graph (cf. the formalization of these four *mqis* in Example 2, specifically M_4–M_7).

5   Evaluation

We implemented an MQI-solver for DECLARE constraints. Our implementation takes as input a DECLARE constraint set C and returns as output MQI(C) and the introduced (quasi) inconsistency measures. We then performed run-time experiments on the following real-life data sets:

– BPI challenge 2017[2]. This data set contains an event log of a loan application process of a Dutch financial institute. The log is constituted of 1,202,267 events corresponding to 31,509 loan application cases.
– BPI challenge 2018[3]. This data set contains an event log of a process at the level of German federal ministries of agriculture and local departments. The log comprises 2,514,266 events corresponding to 43,809 application cases.

[2] https://www.win.tue.nl/bpi/doku.php?id=2017:challenge.
[3] https://www.win.tue.nl/bpi/doku.php?id=2018:challenge.

– Sepsis 2016[4]. This data set contains an event log of a hospital process concerning the treatment of sepsis, which is a life threatening condition. The log contains around 1000 cases with 15,000 events.

We selected these data-sets because they provide recent data from real-life processes. Also, we selected these data-sets to analyze data of domains which are subject to a high degree of regulatory control and sensible to compliant process execution (e.g., financial-, government- and medical sector).

From these logs, we mined declarative process models using Minerful, which is a state-of-the-art tool for declarative constraint discovery [7]. As configuration for mining, we considered the three parameters of *support, confidence* and *interest*. The support threshold indicates the minimum number of traces a constraint has to be fulfilled in for it to be included in the discovered model. Confidence scales the support by the ratio of traces in which the activation occurs, resp. interest scales by the ratio of traces both the constrained tasks occur in. We ran Minerful with a support of 75%, confidence of 12.5% and interest factor of 12.5%, as proposed in the experiment design by [7]. We then applied our implementation to (a) compute all minimal quasi-inconsistent subsets, and (b) compute the $\mathcal{I}_{\mathsf{MI}}^{Q}$ quasi-inconsistency measures, as well as the $\Gamma_{\#}^{Q}$ culpability measures for all constraints. The experiments were run on a machine with 3 GHz Intel Core i7 processor, 16 GB RAM (DDR3) under macOS High Sierra Version 10.13.6.

Table 2. Results of run-time experiments for the analyzed data-sets

Log	BPI Challenge '17	BPI Challenge '18	Sepsis '16
Constraints	305	70	207
$\mathcal{I}_{\mathsf{MI}}^{Q}$ (or # of *mqis*)	28954	25303	7736
Runtime	27074 ms	10930 ms	4379 ms

Table 2 shows an overview of the resp. mined constraints, as well as the number of detected *mqis*, and resp. quasi-inconsistency measures. For the model mined from the BPI'17 log, nearly 29.000 *mqis* were detected. The largest *mqi* had 17 elements. Here, the REG could efficiently be used to detect this subset via a path-based search. In the BPI'18 model, the largest *mqi* contained 22 elements. Also, 62 of the 70 discovered constraints were part of the overall inconsistency (as opposed to only 87/305 constraints in the BPI'17 log). Interestingly, only 70 constraints still lead to a high amount of *mqis*. In the Sepsis'16 log, there were roughly 7.700 *mqis*.

---

[4] https://data.4tu.nl/repository/uuid:915d2bfb-7e84-49ad-a286-dc35f063a460.

Example 6. For illustration, the following shows an actual *mqi* that we detected in the BPI'17 model.

CoEXISTENCE(*A_Accepted, A_Concept*),

CHAINRESPONSE(*A_Concept, W_Validateapplication*).

CHAINRESPONSE(*W_Validateapplication, W_PersonalLoancollection*),

CoEXISTENCE(*A_Accepted, A_CreateApplication*),

RESPONSE(*A_CreateApplication, O_Sent*),

NOTCoEXISTENCE(*W_PersonalLoancollection, O_Sent*)

This actual constraint set returned by the discovery algorithm is quasi-inconsistent. First, *A_Accepted* and *A_Concept* are constrained to appear together. Then, *A_Concept* transitively entails *W_PersonalLoancollection* via two CHAINRESPONSE constraints. Also, *A_Accepted* and *A_CreateApplication* are constrained to appear together. Then, because *A_CreateApplication* occurs, the task *O_Sent* must occur later. However, the last constraint demands that *O_Sent* and *W_PersonalLoancollection* never occur together, both of which are however entailed. In result, this is a quasi-inconsistent subset with the activation *A_Accepted*. Note that the discovery algorithm however did not return a constraint such as PARTICIPATION(*A_Accepted*). Thus, this set of constraints returned by the miner is not inconsistent per se and thus cannot be detected as problematic with existing approaches. Yet, we argue that such a set of constraints should not be contained in any declarative process model, as it is highly confusing and potentially makes the model unusable in practice. Here, our approach allows to detect such problematic sets of constraints as quasi-inconsistent subsets. Table 2 shows that a high number of these *mqis* was actually returned by the miner for all three analyzed logs. As identifying such amount of problematic subsets manually is unfeasible, our approach therefore contributes a feasible means to detect problematic constraints and thus to improve model quality.

In the scope of identifying the actual causes of inconsistency, culpability measures can be used to quantify the degree of blame that individual constraints carry [9]. For the three discovered models, we therefore computed the respective $\Gamma_{\#}^{Q}$ values for all constraints.

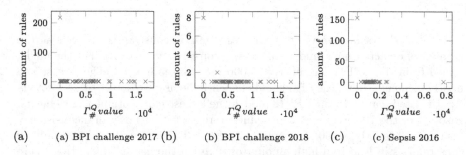

(a)   (a) BPI challenge 2017   (b)   (b) BPI challenge 2018   (c)   (c) Sepsis 2016

Fig. 1. Distribution of culpability values for the constraints in the respective models, using the $\Gamma_{\#}^{Q}$ measure.

Figure 1 shows the distribution of the $\Gamma_{\#}^{Q}$ culpability values for the constraints mined from the respective logs. The x-axis shows the respective culpability value, while the y-axis shows the number of constraints with this value. What can be seen for all analyzed models is that we have a high number of constraints with a culpability value of 0 (i.e. they are not part of any mqi), and only a few number of individual constraints which are highly responsible in the context of the overall inconsistency (i.e. they are part of many mqi). For example, in the constraint set mined from the BPI'2017 log, there are around 200 constraints with a culpability of 0, which can thus be seen as unproblematic. This equates to $\frac{2}{3}$ of all constraints. It is thus possible to identify those (roughly) 100 constraints, which should be attended to. We argue that this is a valuable piece of business intelligence and increases efficiency in managing constraints. Here, the corresponding culpability ranking is a further driver for understanding inconsistencies in the context of resolution strategies. That is, for all the considered models, a few number of constraints can be identified that have the highest culpability values. Thus, these constraints can be strategically targeted first to allow for an effective inconsistency resolution. This is evident in the model mined from the Sepsis 2016 log. There was one specific constraint which was part of all $mqis$, namely RESPONSE($AdmissionNC$, $ReleaseA$). If one would delete this constraint, all quasi-inconsistencies would be resolved. This information could therefore be exploited for effective resolution means. As a further example, the model derived from the BPI'17 log contained the constraint RESPONSE($A\_Incomplete$, $O\_Accepted$), which had the highest $\Gamma_{\#}^{Q}$ value of 16890, meaning one could eliminate over 60% of all $mqis$ while deleting only one constraint.

To summarize, due to the distribution of culpability values, it would be possible to resolve all quasi-inconsistent subsets through targeting selected constraints via the culpability ranking and deleting only these few elements. This would allow to mitigate all *potential* inconsistencies, i.e. implicit inhibition sets, with a low amount of information loss. As mentioned, this is clearly shown for the model of the Sepsis'16 log, where it would be possible to resolve all $mqis$ while deleting only one constraint. In result, we argue that our analysis capabilities by the means of culpability measures provide valuable business insights that can be used as a basis for an informed resolution strategy.

6   Related Work

Our work is related to the discipline of business rules management, i.e., ensuring a consistent set of business rules. In this context, companies have to be supported with means to ensure design-time compliance of declarative process models. While there are some approaches that are aimed to solve problems as discussed in this work by design, i.e. during modelling, this work is related to works that assess an existing set of constraints. This is relevant when existing constraints have to be analyzed, which can be often the case, e.g. analyzing the constraints discovered in process discovery, analyzing a previously modelled set of constraints or analyzing a merged set of constraints after company mergers. A closely related work is that by Di Ciccio et al. [7], who focus on resolving redundancies and inconsistencies in declarative process models. However, as discussed in the introduction, those authors define inconsistency as a model which cannot accept any traces, i.e. it is unsatisfiable. To detect such inconsistencies, those authors represent declarative constraints as finite state automata $\mathcal{A}$, and denote $\mathcal{L}(\mathcal{A})$ as the set of strings accepted by $\mathcal{A}$. Then, those authors can detect inconsistent constraints by identifying those constraint sets that are unsatisfiable via automata products, i.e. $\mathcal{L}(\mathcal{A}') = \emptyset$. As motivated in our introduction, quasi-inconsistent constraint sets can still accept an arbitrary set of traces. Thus, quasi-inconsistency cannot be detected by existing means. Our contribution relative to [7] can be seen in the analysis of the BPI'17 log, which was also analyzed by those authors. Where our approach found nearly 29.000 *potential* inconsistencies, those authors reported 2 inconsistencies. While not inconsistent per se, we argue that quasi-inconsistent sets of constraints such as in $\mathcal{C}_1$ or $\mathcal{C}_2$ should still not be contained in declarative process models, as they can potentially make the model unusable and are highly confusing to modelers. Here, to the best of our knowledge, our approach is the first to offer a tractable solution for detecting all sets of potentially contradictory constraints, i.e. minimal implicit inhibition sets, in declarative process models.

7 Conclusion

In this work, we presented the novel concept of quasi-inconsistencies in declarative process models. As quasi-inconsistencies *potentially* make the model unusable, it is important to detect such problems. Here, we proposed a first approach for such a detection. Through the proposed inconsistency measures, companies are presented with quantitative insights regarding model quality. Element-based culpability measures furthermore allow to prioritize problematic constraints and pin-point individual constraints which should be attended to. Through a computation of MQI based on the reactive entailment graph, our approach is applicable to arbitrary reactive constraints.

Future work could be directed towards the integration of our results, especially the proposed analysis capabilities. In the scope of process discovery, inconsistency measures could be used to as a metric to evaluate the quality of discovered constraint sets. For example, users could define a threshold of allowed quasi-inconsistency. Also, the quantitative insights provided by culpability measures could be used to pin-point the actual causes of quasi-inconsistency, and could thus be integrated into existing methods for resolving errors in declarative process models, e.g. [7], or as a basis for cost-analysis in trace alignment [5].

References

1. Burattin, A., Maggi, F.M., Sperduti, A.: Conformance checking based on multi-perspective declarative process models. Expert Sys. Appl. **65**, 194–211 (2016)
2. Cecconi, A., Di Ciccio, C., De Giacomo, G., Mendling, J.: Interestingness of traces in declarative process mining: the janus LTLp$_f$ approach. In: Weske, M., Montali, M., Weber, I., vom Brocke, J. (eds.) BPM 2018. LNCS, vol. 11080, pp. 121–138. Springer, Cham (2018). https://doi.org/10.1007/978-3-319-98648-7_8
3. Corea, C., Delfmann, P.: Supporting business rule management with inconsistency analysis. In: Proceedings of the BPM 2018 Industry Track (2018)
4. De Giacomo, G., De Masellis, R., Montali, M.: Reasoning on LTL on finite traces: insensitivity to infiniteness. In: AAAI (2014)
5. De Giacomo, G., Maggi, F.M., Marrella, A., Patrizi, F.: On the disruptive effectiveness of automated planning for LTL f-based trace alignment. In: Thirty-First AAAI Conference on Artificial Intelligence (2017)
6. van Der Aalst, W.M., Pesic, M., Schonenberg, H.: Declarative workflows: balancing between flexibility and support. Comput. Sci.-Res. Dev. **23**(2), 99–113 (2009)
7. Di Ciccio, C., Maggi, F.M., Montali, M., Mendling, J.: Resolving inconsistencies and redundancies in declarative process models. Inf. Syst. **64**, 425–446 (2017)
8. Dwyer, M.B., Avrunin, G.S., Corbett, J.C.: Patterns in property specifications for finite-state verification. In: Proceedings of the 21st International Conference on Software Engineering (1999)
9. Grant, J., Hunter, A.: Measuring consistency gain and information loss in stepwise inconsistency resolution. In: Liu, W. (ed.) ECSQARU 2011. LNCS (LNAI), vol. 6717, pp. 362–373. Springer, Heidelberg (2011). https://doi.org/10.1007/978-3-642-22152-1_31
10. Grant, J., Martinez, M.V.: Measuring Inconsistency in Information. College Publications, London (2018)
11. Hunter, A., Konieczny, S., et al.: Measuring inconsistency through minimal inconsistent sets. KR **8**, 358–366 (2008)
12. Maggi, F.M., Di Ciccio, C., Di Francescomarino, C., Kala, T.: Parallel algorithms for the automated discovery of declarative process models. Inf. Syst. **74**, 136–152 (2018)
13. Markey, N.: Past is for free: on the complexity of verifying linear temporal properties with past. Acta Informatica **40**(6–7), 431–458 (2004)
14. Pesic, M.: Constraint-based workflow management systems: shifting control to users (2008)
15. Sadiq, S., Governatori, G.: Managing regulatory compliance in business processes. In: vom Brocke, J., Rosemann, M. (eds.) Handbook on Business Process Management 2. IHIS, pp. 265–288. Springer, Heidelberg (2015). https://doi.org/10.1007/978-3-642-45103-4_11
16. Thimm, M.: On the expressivity of inconsistency measures. Artif. Intell. **234**, 120–151 (2016)
17. Thimm, M.: On the evaluation of inconsistency measures. In: Grant, J., Martinez, M.V. (eds.) Measuring Inconsistency in Information. Studies in Logic, vol. 73. College Publications, London (2018)

Decision Support for Declarative Artifact-Centric Process Models

Simon Voorberg[✉], Rik Eshuis, Willem van Jaarsveld,
and Geert-Jan van Houtum

Eindhoven University of Technology, Eindhoven, The Netherlands
{s.voorberg,h.eshuis,w.l.v.jaarsveld,g.j.v.houtum}@tue.nl

Abstract. Data-driven business processes involve knowledge workers that process information to take decisions. Such processes have been modelled successfully using artifact-centric process models. Artifacts represent business entities about which the knowledge workers collect and process information. Since information retrieval costs time and money, the key goal is to retrieve only the pieces of information that are needed to make a well-informed decision. To aid knowledge workers in achieving this goal, this paper realizes decision support for declarative artifact-centric process models by showing how declarative artifact-centric process models can be translated into Markov Decision Processes (MDP). The approach is illustrated with an example from the field of financial services.

1 Introduction

Knowledge workers are responsible for decision making and analyzing information in the Knowledge-intensive Processes they perform [7,16]. Decisions for cases (process instances) are made in the context of business entities such as loan requests, price quotes or maintenance orders. Progress in such decision-intensive processes relies on available information. Knowledge workers prepare decisions by performing tasks in which information is gathered. The abundance of data in combination with the rise of data analytics techniques has increased the amount of information that is potentially available.

A knowledge worker typically has the discretionary power to perform or skip an information-gathering task, which requires a decision from his side. While performing an information-gathering task typically improves the quality of the final decision, it also increases the throughput time of the process and the costs made to reach the decision. For instance, a mortgage expert can gain more in-depth information about the client by assessing her/his risk level, but doing so may take a day and involve costs for getting a report from an outside organization. Therefore, knowledge workers need to continually make trade offs whether they need more information or whether they have sufficient information to make a reliable decision (cf. Fig. 1). Advanced support for such decision-intensive processes is currently lacking in scientific literature [12].

© Springer Nature Switzerland AG 2019
T. Hildebrandt et al. (Eds.): BPM 2019, LNBIP 360, pp. 36–52, 2019.
https://doi.org/10.1007/978-3-030-26643-1_3

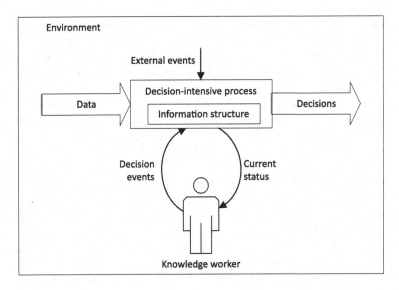

Fig. 1. Decision-intensive processes

We introduce decision support for knowledge workers based on *Artifact-centric process models*, a technique that combines data and process aspects in a holistic way in order to model decision-intensive processes [17,24]. In particular, we focus on declarative business artifact-centric process models, which offer additional flexibility in performing processes by specifying constraints as rules rather than sequence relations and by allowing external events to influence the process. We model the artifact-centric process using the Guard-Stage-Milestone (GSM) formalism [5,13]. GSM is well-defined and is one of the base models that have been used to introduce the OMG Case Management Model and Notation (CMMN) [3]. We slightly extend GSM by introducing discretionary decision events to support the approach.

In GSM schemas for decision-intensive processes [24], information is gathered and computed, i.e., data attributes are written. Initially, only a few attributes are known, such as the name of the client; other attributes such as salary and age are initially unknown, but estimates can be derived from previous mortgage requests. The main issue for knowledge workers is to decide whether they have collected sufficient information to make the final decision, or whether they need to retrieve more information, i.e., replace initial estimates with real values paying a certain cost, to improve the quality of the final decision. Additionally, if more information is needed, then the knowledge worker needs to decide which data attribute(s) to retrieve next.

To guide the knowledge worker in this decision, we introduce the novel concept of an *information structure* for decision-intensive processes. An information structure estimates the retrieved level of information, denoted as the *quantity of interest*, based on retrieved sources of information and estimates for non-retrieved information. For instance, an information structure for mortgages could estimate the creditworthiness of the clients. This estimate is updated as more information is retrieved about the client. The information structure is used to specify the recommended outcome of the final decision, but the actual outcome is decided by the knowledge worker.

To support decision making in declarative artifact-centric processes, we use the probabilistic optimization model: *Markov Decision Process* (MDP) [20]. This model recommends the best decision based on the information that is currently available in an environment where everything else is uncertain and only probabilities are known. It is uncertain what the real outcome of a decision is, but we can do predictions based on the probability distributions of the possible future decisions. We contribute by defining an approach for mapping a GSM schema to an MDP. In the MDP, the structure of the artifact process is incorporated. The information structure is used to define conditions on when the MDP should be terminated. By modelling the uncertainty explicitly with probabilities, we improve the reliability of the decision support and offer user guidance compared to other techniques such as simulations [21] and fuzzy modeling [8], as we can derive expected cost based on these probabilities.

The remainder of this paper is structured as follows. Section 2 gives preliminaries on GSM schemas and MDPs. Section 3 introduces the notion of the information structure. Section 4 discusses the mapping from GSM schemas to MDPs. Section 5 discusses related work. Section 6 discusses the approach, concludes the paper and gives future research topics. Due to space limitations, the formal translation from GSM schemas to MDPs is not provided here, but in an online appendix [26].

2   Preliminaries

2.1   Guard-Stage-Milestone Schemas

In this section, we formally define the GSM schemas used in this paper. We consider a lightweight variant compared to the classical GSM schemas [5,13]. In particular, the GSM schemas in this paper are without hierarchy and have monotonic executions [9]. Using such a lightweight GSM schema variant allows us to better highlight the key aspects of the translation. In future work, we can relax these assumptions as, for example, adding hierarchy is orthogonal to the developed decision support.

We informally introduce GSM schemas by means of an example in Fig. 2. Rounded rectangles represent stages, in which work is performed. Open circles denote milestones, which are business objectives typically achieved by performing work in stages. Diamonds indicate sentries, which are rules that specify when a stage is opened. Milestones also have sentries that specify when milestones are achieved, but these are not visualized. Table 1 contains the sentries (guards) for the stages and the sentries of the milestones of Γ.

Departing from original GSM notation [13], we explicitly visualize external events using the bull's eye symbol. Named external events (prefix E:) are generated by the environment and not under the control of the knowledge worker performing this process, while decision events (prefix D:) are controlled and generated by the knowledge worker. Rectangles denote data attributes, written in the stages to which they are connected.

The GSM schema models a mortgage process. Input is a request for the issuing of a mortgage with a certain value to a customer. A mortgage expert decides whether or not a mortgage of a certain amount is issued. To prepare a decision, the expert can gather information by retrieving a set of criteria that predict the customer's creditworthiness. In this case, we consider a set of four criteria: salary, outstanding debts, employment contract and age. The mortgage expert decides which criteria to check in which order and also decides on the final outcome of the process. Once the mortgage request arrives, the salary is immediately checked based on a process constraint. After this process step has been completed, the other three stages can be opened upon discretion of the mortgage expert. The 'Make Decision' stage can be opened at any time, once the mortgage expert has collected sufficient information and can make a decision to accept or reject this mortgage request.

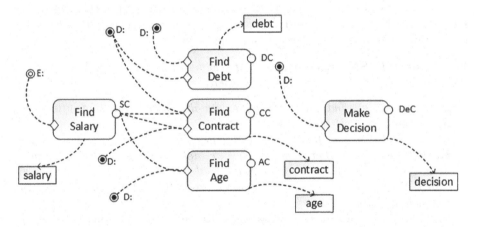

Fig. 2. Mortgage decision process

To keep track of all business-relevant information about an artifact instance as it moves through its lifecycle, GSM schemas use data attributes and status attributes. Data attributes store information on an artifact instance and can be of any data type, for instance, integer or string. In our decision making process we assume that a task in an atomic stage yields new information that is stored in one or more data attribute(s). The status attributes are boolean attributes that keep track of the GSM lifecycle of the artifact instance; a stage (milestone) is true when it is open (has been achieved).

Table 1. Sentries for the Mortgage decision process

Stage	Sentry (Guard)
Find Salary	E:MortgageRequest
Find Debt	D:Debt ∧ not(Make Decision) ;
	D:Debt+Contract ∧ not(Make Decision)
Find Contract	D:Contract ∧ SC ∧ not(Make Decision) ;
	D:Debt,Contract ∧ SC ∧ not(Make Decision)
Find Age	D:Age ∧ SC ∧ not(Make Decision)
Make Decision	D:Decision

(a) Stage sentries

Milestone	Sentry
SC	C:Find Salary
DC	C:Find Debt
CC	C:Find Contract
AC	C:Find Age
DeC	C:Make Decision

(b) Milestone sentries

Each sentry is triggered by an event and has a condition that references attributes of the GSM schema. We assume a condition language C that includes predicates over scalars and boolean connectives between all attributes in the model. In the event part, we distinguish between external events (prefix E:), decision events (prefix D:) and completion events (prefix C:). External events come from the environment of the process, typically a customer, but not the knowledge worker. Decision events are generated by knowledge workers to start a discretionary activity. Knowledge workers have the freedom and authority to perform or skip such an activity. Decision events are not used in classical GSM schemas [13], but are introduced here to model discretionary activities. Finally, completion events signal completion of atomic stages.

Definition 1 (GSM schema). *A GSM schema is a tuple $\Gamma = (\mathcal{A} = \mathcal{D} \cup \mathcal{S} \cup \mathcal{M}, wt, time, \mathcal{E} = \mathcal{E}_{cmp} \cup \mathcal{E}_{ext} \cup \mathcal{E}_{dec}, \mathcal{R})$, where*

- *$\mathcal{A}$ is the set of attributes containing the data attributes $\mathcal{D}$, the stage attributes $\mathcal{S}$, and milestone attributes $\mathcal{M}$;*
- *$wt : \mathcal{S} \to \mathcal{P}(\mathcal{D})$ is a function that specifies for each stage the set of data attributes written in that stage. We require that distinct stages write distinct variables, i.e., for $s, s' \in \mathcal{S}$, if $s \neq s'$ then $wt(s) \cap wt(s') = \emptyset$.*
- *time : $\mathcal{S} \to \mathbb{N}$ is a function that assigns to each stage the time needed to complete the stage;*
- *$\mathcal{E}$ is a finite set of external events, consisting of named external events $\mathcal{E}_{ext}$ (prefixed E:), completion events $\mathcal{E}_{cmp} = \{C:s \mid s \in \mathcal{S}\}$, and decision or discretionary events $\mathcal{E}_{dec}$ (prefixed D:);*
- *$\mathcal{R}$ is a function from $\mathcal{S} \cup \mathcal{M}$ to a set of sentries (see Definition 2) ranging over all attributes $\mathcal{A}$ defined on the condition language C.*

Definition 2 (Sentry). *A sentry has the form $\tau \wedge \gamma$, where τ is the event-part and γ the condition-part. The event-part τ is either empty (trivially true), an external event $e \in \mathcal{E}$ or an internal event $+d$ (d becomes true) or $-d$ (d becomes false), where $d \in \mathcal{S} \cup \mathcal{M}$ is a stage or milestone attribute. The condition γ is a Boolean formula in the condition language C that refers to $\mathcal{A}$, so data attributes in $\mathcal{D}$ and status attributes in $\mathcal{S} \cup \mathcal{M}$. The condition-part can be omitted if it is equivalent to true.*

We also introduce the auxiliary function

$$tr : \mathcal{E}_{dec} \to \mathcal{P}(\mathcal{S})$$

where $tr(\text{D}:n) = \{ s \in \mathcal{S} \mid \text{D}:n$ is trigger event of a sentry of $s \}$.

At any given point of time, the whole GSM schema is in a specific state, called a snapshot, that is defined by the values of the status attributes and data attributes.

Definition 3 (Snapshot). *For a GSM schema $\Gamma = (\mathcal{A}, wt, time, \mathcal{E}, \mathcal{R})$ a snapshot is a mapping σ from the attributes in $\mathcal{A}$ into appropriate attribute values. Initially, all data attributes have value $\bot$ (unknown) and all stage and milestone attributes have value False.*

In response to the occurrence of an event in $\mathcal{E}$, a snapshot changes into another snapshot by performing a Business step [5,9]. The event can result in sentries that evaluate to true, which in turn may lead to stages being opened or closed and milestones being achieved. In particular, a stage completion event signals that a task has been completed; the payload of the event carries new values for the data attributes written in the stage. These values are incorporated into the new snapshot.

An important assumption in this variant of GSM schemas is that completion events happen after a known time period and are not used to open new stages, i.e, they are not used in guards. This means that we know beforehand how long a certain action will take, so the time of completion can not affect the decisions. This assumption is made to keep the later introduced MDP translation understandable, but will be relaxed in future work.

2.2   Markov Decision Process (MDP)

In this section we introduce the semantics that are used to define an MDP model and give some intuition on how to solve these models. MDPs are used to model dynamic processes in which repeated decisions (also: actions) are taken facing uncertainty. Decisions are made in each *decision epoch* $t \in \mathbf{N}$ based on the *state* $st(t)$ of the system. The state contains the information available in the epoch; S denotes the set of all possible states. The goal is to decide which action to take from the set of allowed actions $A(st)$ in state st, in order to minimize a given cost function. The action determines the direct costs incurred, and influences the next state but does not completely specify it:

Definition 4 (Cost function). *$C_a(st)$ is the cost when starting from state st, and doing action a. In our model this cost does not depend on the next state.*

Definition 5 (Transition Probability). *$P_a(st, st')$ is the probability that, after taking action a in state st, we end up in state st'.*

The complete action space is denoted by $\mathbb{A} = \cup_{st \in S} A(st)$. An MDP is then a 4-tuple:

Definition 6 (Markov Decision Process). *A Markov Decision Process is a tuple* $Z = (S, \{A(st)\}_{st \in S}, \{P_a(st, st')\}_{st,st' \in S, a \in A(st)}, \{C_a(st)\}_{st \in S, a \in A(st)})$.

For the MDPs considered in this paper, states that correspond to situations where the final decision was made are so-called terminating states. After reaching such states, costs are no longer incurred. So effectively, the goal is to minimize the total costs incurred until a terminating state can be reached: Our MDP has an infinite number of decision epochs only for mathematical convenience.

To minimize total costs, we aim at reaching one of the terminating states as fast as possible: It may be possible to terminate while only a subset of all possible actions were performed, and this depends crucially on the outcomes of these actions. Our approach estimates the action that yields the cheapest route to a terminating state. In particular, starting in epoch T, the approach chooses $a(t)$, the action at time epoch t, for all $t \geq T$, to minimize $V = \mathbb{E}[\sum_{t=T}^{\infty} C_{a(t)}(st(t))]$. This can be achieved by solving the following set of recursive equations, for all states $st \in S$, starting from the terminating states [20]:

$$V^t(st) = \min_{a \in A(st)} C_a(st) + \sum_{st' \in S} P_a(st, st')(V^{t+1}(st'))$$

A range of algorithms exist to solve the above problem and give the statistically most interesting action to do, based on the current state [20]. These algorithms should suffice for modestly sized schemas. This paper focuses on translating GSM models into MDPs and leaves a detailed investigation into solving these MDPs (i.e. selecting optimal actions) for future research.

3 Information Structure

The need for an information structure mainly arises because the GSM schema has no single variable that can summarize the current snapshot.

To arrive at such a variable, we note that attributes retrieved in knowledge-intensive processes by definition help to obtain information about a specific property associated with each case. We will refer to this property as the *quantity of interest*. For example, in a mortgage request the quantity of interest may be the creditworthiness of the client.

3.1 Goal

We need a mathematical model to be able to create this variable and define the GSM snapshots where enough information is known to make a reliable final decision.

Definition 7 (Information structure). *Given a set of data attributes* $\mathcal{D} = \{X_1, X_2, ...X_k\}$, *a subset of* $\mathcal{D}$ *with the retrieved values at snapshot* σ: $\sigma(\mathcal{D}) = \{X_1 = x_1, X_4 = x_4, ...\}$ *and probability distributions* $F_{X_j | X_j \in \mathcal{D} \setminus \sigma(\mathcal{D})}$ *of the non-retrieved values, the information structure is a function* $\hat{y} = f(\sigma(\mathcal{D}), F_{X_j})$, *which summarizes the known information about the quantity of interest* y.

We emphasize that the outcome of the information structure is always an estimate of the real value of y. Therefore, the estimated y is denoted by $\hat{y}$.

A wide range of methods exist that can yield information structures from past cases, e.g. (non)linear regressions, neural networks, etc. [11, 19].

3.2   Linear Regression

To estimate the correct weighing of different variables and create an information structure, linear regression has shown to be an effective method and is therefore used as an example method for the information structure. It uses previous cases to estimate the weights of each individual variable by minimizing the sum of squared errors. This preliminary set consists of two parts. First, the y_i variables are the dependent variables for historical case i, in our example this is the eventual creditworthiness. Second, the $x_{i,j}$ variable is the jth independent variable in case i. Formally, linear regression is then defined as:

Definition 8 (Linear regression). *Given a set of data variables $\{y_i, x_{i,1},$ $x_{i,2}, ..., x_{i,k}\}_{i=1}^n$, we define the linear regression as $y_i = \beta_0 + \beta_1 x_{i,1} + \beta_2 x_{i,2} + ... + \epsilon_i$, where ϵ_i denotes the regression error. By minimizing the sum of the squared errors over all n estimates of y, we obtain the best β values: $\arg min_\beta \sum_{i=1}^n (y_i - \beta_0 - \sum_{j=1}^k \beta_j x_{i,j})^2$.*

We remark that this method has a set of assumptions over the set of data variables that need to be checked before the estimates are reliable [23]. An example is the need for no or little multicollinearity in the data. The regression error ϵ_i incorporates certain factors that might not have been recognized as relevant sources of information. Based on the β values or weights that are derived in the linear regression, we are now able to build a function that can be used as information structure. Given a set of variables or attributes that were retrieved, $\{x_1, x_2, ..., x_k\}$, we can define:

$$\hat{y} = \beta_0 + \beta_1 x_1 + \beta_2 x_2 + ... + \beta_k x_k$$

Returning to our example, suppose we have a set of attribute values for a customer:

$$\sigma(\mathcal{D}) = \sigma(\text{salary}, \text{debt}, \text{contract}, \text{age}, \text{decision}) = \{20000, 0, 1, 26, \bot\}$$

where the last attribute is used to store the final decision and thus depends on the creditworthiness y. Also, $\text{contract} = \begin{cases} 1 & \text{if fixed} \\ -1 & \text{if temporary} \end{cases}$.

Using all previous cases of mortgage issuing to estimate the betas, we get the information structure: $\hat{y} = \beta_0 + \beta_{salary}\text{salary} + \beta_{debt}\text{debt} + ...$

Suppose Table 2 gives the β values that were determined using linear regression. Based on this, the creditworthiness of this customer is estimated at a level of

$$\hat{y} = \beta_0 + \beta_{salary}\text{salary} + \beta_{debt}\text{debt} + \beta_{contract}\text{contract} + \beta_{age}\text{age}$$
$$\hat{y} = 2 + 0.5 + 0 + 1 + 0.325 = 3.825$$

Table 2. Beta values

Beta	Value
β_0	2
β_{salary}	0.000025
β_{debt}	−0.005
β_{age}	0.0125
$\beta_{contract}$	1

3.3 Estimating y with Uncertain Variables

In the case of a GSM schema where decisions need to be made during the process, the problem is that not all attribute values are known yet. One needs to work with estimates based on the probability distribution of the attributes. This probability distribution can for example be obtained from previous cases. We want to estimate $\hat{y}$ by using the probability distributions of the other variables. The decision rule could, for instance, be defined as: accept if $P(\hat{y} \geq \theta) \geq \phi$. Although the value of some attributes might be unknown, the regression estimate of the betas is still valid. Therefore, using the linear function of the value of interest, we can still estimate the outcome.

Returning to our example, suppose that the salary, debts and age are already retrieved, but the type of employment contract is unknown. Furthermore, historically 60% of employees has a fixed contract, the threshold for the mortgage request is set at $\theta =$ mortgage amount$/50000$, and $\phi = 0.5$. Using the β estimates from Table 2, we can now estimate the real value of the customer with a 150000 mortgage request ($\theta = 3$).

$$P(\hat{y} \geq \theta | \mathsf{salary} = 20000, \mathsf{debt} = 0, \mathsf{age} = 26)$$
$$= P(\beta_0 + \beta_{salary}20000 + \beta_{debt}0 - \beta_{age}26 + \mathsf{contract} \geq 3)$$
$$= P(\mathsf{contract} \geq 0.175)$$

Because contract can only take values 1 and −1 and the probability of a fixed contract was said to be 0.6, this results in $P(\hat{y} \geq 3|st) = 0.6$. Therefore, since the rule is adopted that $P(\hat{y} \geq \theta) > 0.5$ suffices to accept the request, the knowledge worker could now be advised to abort the retrieval of contract information and to accept the request. (Never mind that banks are in reality typically not so forthcoming, and would likely use higher values of ϕ.)

4   Decision Support for GSM Schemas with Uncertainty

In this section we discuss the implementation of MDP support in a GSM environment. First, we introduce a framework that links the two different concepts and gives an idea of how the new process would include decision support. Second, we introduce a more formal explanation of translating the GSM notation to the

MDP model. The translation is formally introduced in an online appendix [26]. Finally, based on the introduced problem of deciding on a mortgage request, we give an example of how the process could run.

4.1 Framework

Figure 3 gives a framework of our solution approach. This figure only shows the GSM components that have an effect on the MDP model and are used when creating a decision event. The GSM part of this figure works according to the earlier explained rules, where an instance of the schema is in a current snapshot. By performing business steps we open and close stages, so we end up in new snapshots. Suppose we are in a snapshot where one of the possible actions is a decision event. This means that knowledge workers now have to decide if they should initiate this event. However, the knowledge worker asks for support. Therefore, using the information structure, the MDP model tries to estimate the effects of the different options that are available. The information structure helps with translating the snapshot of a GSM schema into a variable that indicates the progress of the total process. Based on this and the MDP state, the set of possible actions is determined. By using an MDP optimization algorithm, we can now come up with an advice on what to do next. Subsequently, the knowledge worker has the freedom to decide using the support. Based on the decision a new event will happen, causing a new Business step to happen. As a consequence, we receive a new snapshot and the process repeats.

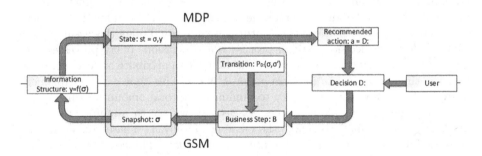

Fig. 3. GSM to MDP mapping

4.2 Translation GSM to MDP

To translate the GSM lifecycle into an MDP state set, we recall the data snapshot $\sigma(\mathcal{D})$, which consists only of the data attributes. A state in the MDP can be compared to a data snapshot in the GSM schema containing all retrieved information. A state change in the MDP model implies the revealing of new data attributes through a business step in the GSM schema in which the stage completes that writes the data attribute. The new value of this attribute changes $\sigma(\mathcal{D})$ and thus also the state of the MDP. We assume that the current status of stages and milestones does not influence the decision process of a knowledge worker; we plan to relax this assumption in future work.

As the basic goal of an MDP is to decide what action to perform based on the minimization of costs, we need to recognize possible actions in the GSM schema. To jump from one snapshot to another in GSM, we take a business step that is deployed by an arriving event. More specific, the set of decision events, $\mathcal{E}_{dec}$, defines the possible actions to consider as the knowledge worker is only allowed to create these events: $a \in \mathcal{E}_{dec}$. If we look at the GSM notation, $tr(a) \subset \mathcal{S}$ is the set of opened stages for action a and $wt(tr(a)) \subset \mathcal{D}$ is the set of retrieved attributes for action a.

The result of an action is unknown in the MDP model. Comparably, in the GSM schema, the decision to collect information on a customer yields an unknown answer. In the MDP schema, we model this uncertainty using the transition probability to a specific state. Therefore, for each unknown attribute that can be determined in the GSM schema (except the final decision attribute), we need to obtain a probability distribution F_{X_j}. If we recall the set of historical information, $\{y_i, x_{i,1}, x_{i,2}, ..., x_{i,k}\}_{i=1}^n$, which was used in Sect. 3, we can use this information to define the empirical probabilities of ending up in a certain state.

$$P(\sigma(X_j) = x_j) = \frac{\sum_{i=1}^n I(x_{i,j} = x_j)}{n}, \text{ where } I(x_{i,j} = x_j) = \begin{cases} 1 & \text{if } x_{i,j} = x_j \\ 0 & \text{otherwise} \end{cases}$$

By assuming independence between attributes, we can take the product of individual attribute probabilities to find the transition probabilities for the retrieval of multiple attributes in one action. Future research could relax this assumption.

Considering a decision-making process where decisions delay processes, we assign the necessary time to determine the value of an attribute as cost. Suppose that the time necessary to perform action a is C_a. This time can be explained in two different ways. First, we try to minimize the total amount of time spent by workers on the complete process, T. Second, we try to minimize the time until a decision is made to make the customer happy, τ. Therefore, we define the cost function as a combination of these two goals. The cost for a is $C_a = \delta T(a) + (1 - \delta)\tau(a)$, where δ is the parameter indicating the relative importance between the total time and the customer waiting time. The cost for parallel stages in one action can be defined as: $T(a) = \sum_{s \in tr(a)} T_s$. The waiting time of this action is equal to: $\tau(a) = \max_{st \in tr(a)} T_s$. T_s is the time needed for stage $s \in \mathcal{S}$: $time(s)$. By opening stages in parallel, the retrieval is less time-consuming and thus can be an advantage. The exception for this given cost function are the cost to make a final decision.

Using the resulting variable $\hat{y}$ from the information structure, we can define a stopping rule for the MDP. In the case of our example this could be: accept if $P(\hat{y} \geq \theta | \sigma(\mathcal{D})) \geq \phi$. Here, we test if the variable $\hat{y}$ is greater than θ with a probability of at least ϕ, given that we are in state $\sigma(\mathcal{D})$. If, based on the information of the current state σ, this rule will hold, a knowledge worker could make the decision to abort any further information retrieval because the $C_{D:Accept}$ to terminate are 0. Using the information structure, the stopping rule and the probability distributions, it is possible to calculate for each stage in the GSM model if a stopping rule has been achieved.

4.3   Example

We return to the example that was introduced in Sect. 2.1, and illustrate the mapping by naming a few examples. Suppose at some point only the attribute salary is retrieved, and that its value is 20000. Then the state can be represented as:

$$\sigma(\mathcal{D}) = \sigma(\text{salary, debt, contract, age, decision}) = \{20000, \bot, \bot, \bot, \bot\}$$

The action space for state $\sigma(\mathcal{D}) = \{20000, \bot, \bot, \bot, \bot\}$ is:

$$A_\Gamma(\sigma(\mathcal{D})) = \{\{\text{D:Age}\}, \{\text{D:Debt,Contract}\}, \{\text{D:Debt}\}, \{\text{D:Contract}\},$$

$$\{\text{D:Accept}\}, \{\text{D:Decline}\}\}$$

Now suppose we choose to retrieve the debt attribute, and in 40% of previous cases the debt amount was found to be 0. Then we can find the following transition probability:

$$P_{D:Debt}(\{20000, \bot, \bot, \bot, \bot\}, \{20000, 0, \bot, \bot, \bot\}) = 0.4$$

Furthermore, if 60% of previous cases has a fixed contract, then

$$P_{D:Contract}(\{20000, \bot, \bot, \bot, \bot\}, \{20000, \bot, 1, \bot, \bot\}) = 0.6$$

$$P_{D:Debt,Contract}(\{20000, \bot, \bot, \bot, \bot\}, \{20000, 0, 1, \bot, \bot\}) = P_{D:Debt} \cdot P_{D:Contract} = 0.24$$

As for the cost function, suppose $\delta = 0.5$, $time(\text{Debt}) = 200$, $time(\text{Contract}) = 150$. Then:

$$C_{D:Debt,Contract}(\{20000, \bot, \bot, \bot, \bot\}) = 0.5T(a) + 0.5\tau(a) =$$

$$0.5(time(\text{Debt}) + time(\text{Contract})) + 0.5\max(time(\text{Debt}), time(\text{Contract}))$$

$$= 0.5(200 + 150) + 0.5 \cdot 200 = 275$$

Also, $C_{D:Debt}(\{20000, \bot, \bot, \bot, \bot\}) = 200$ and $C_{D:Contract}(\{20000, \bot, \bot, \bot, \bot\}) = 150$. The MDP weighs the relative merits of the various possible actions. E.g. retrieving only debt may yield enough information to decline the request, but if the value of debt retrieved does not lead to the ability to make a final decision, then contract is to be retrieved after all. If the latter scenario is likely, then retrieving both *simultaneously* is advisable, since this is less costly (faster) then one-by-one retrieval. Conceptually, the MDP approach weighs the eventual expected outcome of all alternatives actions, and recommends the action that has the most favorable expected outcome. Based on this, suppose we take the action D : Debt,Contract, then we can possibly end up in state: $\{20000, 0, 1, \perp, \perp\}$.

Now, in general, the decline action moves us directly to the state where the final decision is decline:

$$P_{D:\text{Decline}}(\{20000, 0, 1, \perp, \perp\}, \{20000, 0, 1, \perp, \text{decline}\}) = 1$$

To make the final decision of declining the request, cost are:

$$C_{D:\text{Decline}}(\{20000, 0, 1, \perp, \perp\}) = \begin{cases} \infty & \text{if } P(\hat{y} \leq \theta_{dec}|\text{salary, debt, contract}) \leq \phi_{dec} \\ 0 & \text{if } P(\hat{y} \leq \theta_{dec}|\text{salary, debt, contract}) \geq \phi_{dec} \end{cases}$$

Suppose that $\hat{y}$ is the linear regression according to the example in Sect. 3, and that $\theta_{dec} = 4$ and $\phi_{dec} = 0.6$. Then

$$P(\hat{y} \leq 4|\text{salary} = 20000, \text{debt} = 0, \text{contract} = 1) = P(\beta_{age}\text{age} \leq 0.5) = P(\text{age} \leq 40)$$

Assuming that 65% of all requesters are younger than 40, $C_{D:\text{Decline}}(\{20000, 0, 1, \perp, \perp\}) = 0$. The result is that the cheapest option of actions is to decline and thus the recommendation will be to decline. In all cases where the stopping rule is not fulfilled, $C_{D:\text{Decline}} = \infty$, causing actions where more information is retrieved to always be cheaper. E.g., suppose that instead of the above assumption only 20% of requesters are younger than 40. Then rejecting the case without retrieving the age would be too risky. Instead, the advice would be to retrieve the age of the requester, which would allow the final decision to be made.

5   Related Work

In the area of business process modeling there has been some related work on using Markov Decision Processes to give decision support. Vanderfeesten et al. [25] introduced Markov Decision Processes for optimizing the execution of Product Data Models, which specify data elements and operations that transform data elements into other data elements. Uncertainty in their approach is about the success or failure of operations, whereas in our approach uncertainty is about the quantity of interest that needs to be decided upon.

The paper of Petrusel [18] extends the paper of Vanderfeesten et al. [25]. Using a more expanded version of the Product Data Model, the Decision Data Model (DDM), they use MDP models to make optimal decisions in this DDM. The contribution we make compared to these papers [18,25] is taking into account the actual data values that were retrieved during the process, whereas the previous papers only considered data elements without values. MDPs with structural similarities to the MDPs that we retrieve from GSM have been studied before by Lim, Bearden and Smith [15], but not in the context of business processes. They focus on the search of attributes to discover the value of an option, without any process restrictions.

Schonenberg et al. [22] focus on giving recommendations based on the comparison with similar traces in historical cases of the process. We use the historical cases to estimate the probability distributions. Lakshmanan et al. [14] use an instance-specific Probabilistic Process Model (PPM), to define the transition probabilities of a Markov Chain. However, they only give likelihood estimations of the future, but we provide a recommendation based on minimizing future cost.

Ghattas et al. [10] not only focuses on control flow decisions, but also decisions embedded in an action. They use a learning algorithm comparing historical cases to the current case. Our model is more detailed as it introduces an information structure that helps in deciding when to end the entire process and also gives a recommendation on what final decision should be made.

Mertens et al. [16] introduce a new declarative process language: DeciClare, which is an alternative language for modeling decision-intensive processes. The data perspective is based on the Decision Model and Notation [6], which is an industry-standard for modeling the requirements and logic of business decisions. However, DMN and therefore DeciClare do not consider uncertainty regarding the quality of interest to be decided upon, nor do they provide any recommendation support to guide knowledge workers.

Eshuis and Firat [8] use fuzzy modeling to express uncertainty in a more qualitative way. But especially in processes with a high repetitiveness, it is helpful to use the available information of historical cases. Therefore, we make use of the quantitative approach with probabilities. A different approach to model future uncertainty is by doing simulation [21]. However, the flexibility to have unknown decisions and to allow the decision maker to make counter-intuitive decisions is hard to model by simulation as this requires human interaction. Barba et al. [1] use a constrained - process approach, where mainly control flow and resources are considered and decisions are made to optimally divide work over the resources. All these approaches do not make use of quantitative uncertainty by means of probabilities, such as we introduced in this paper.

Conforti et al. [4] use a method to predict the risk of taking certain decisions. Based on this risk, the knowledge worker is recommended the next task. [14] provides likelihood estimates of the future states of the process, based on the Markov process that is estimated. Furthermore, [22] do not give direct recommendations on what action to perform. They only give a do or don't advice per option based on process mining logs. Batoulis et al. [2] use a Bayesian Network to define the dependencies and define an influence diagram. Based on the influence diagram a decision model is defined using DMN. In our model we give more specific recommendations to increase the relevance of the recommendation to the knowledge worker. Moreover, we allow the knowledge worker to make stubborn decisions and include the result of this decision to recalibrate our recommendation.

6   Discussion and Conclusion

In this paper we have introduced a new approach of giving decision support for declarative artifact-centric processes. We do this by translating GSM schemas into a Markov Decision Process, using the novel notion of an information structure that estimates the quantity of interest. The introduced solution can be used as support for many sorts of decision-intensive processes, where the knowledge worker can decide on retrieving different sources of information to gain more knowledge about a case and finally makes a decision using this information.

In order to define a simple translation, we considered GSM schemas without hierarchy. The inclusion of hierarchy complicates our translation as the decision to open certain non-atomic stages would only have indirect consequences for the retrieval of data attributes. This would lead to dependencies between different stages, where all stages are now assumed to be independent of each other. A similar reasoning prevents us now from using non-monotonic executions. Once we have introduced attribute dependencies in the information structure and MDP, both these assumptions can be relaxed. A second assumption in this paper is that we assume that all stages can be decided to be opened. External events or completion events do not trigger the opening of new stages. This allows us to estimate the time needed to perform tasks as we assumed in Sect. 4. The authors will allow for other events than decisions in a follow-up paper.

Also, for the MDP that appears from these translations, there are currently some drawbacks. Firstly, by including all possible values of all variables in the state space, this state space explodes already for very small models. An exploding state space makes it hard or sometimes impossible to find globally optimal solutions. As mentioned in Sect. 2.2, this problem is left for future research. Secondly, the transition probabilities are now modelled as the product of independent attribute probabilities. But many variables in these process are often correlated and cannot be assumed to be independent. In future research we plan to resolve this issue. Finally, the information structure can be modelled using much more complicated models, as was mentioned in Sect. 3, that were not discussed in this paper.

This paper introduces a new approach to give decision support to knowledge workers performing decision-intensive approaches. There are several directions for future research to further validate, refine and improve the approach. One direction is to consider more general GSM schemas and to consider techniques to solve the MDP models in an efficient manner. Another more practical direction is to implement the translation in a tool and apply it in several real-world case studies. In developing such a tool, we will also plan to explore different ways to deal with the state space explosion problem.

References

1. Barba, I., Weber, B., Del Valle, C., Jiménez-Ramírez, A.: User recommendations for the optimized execution of business processes. Data Knowl. Eng. **86**, 61–84 (2013)
2. Batoulis, K., Baumgraß, A., Herzberg, N., Weske, M.: Enabling dynamic decision making in business processes with DMN. In: Reichert, M., Reijers, H.A. (eds.) BPM 2015. LNBIP, vol. 256, pp. 418–431. Springer, Cham (2016). https://doi.org/10.1007/978-3-319-42887-1_34

3. BizAgi, et al.: Case Management Model and Notation (CMMN), v1.1 (2016), OMG Document Number formal/2016-12-01, Object Management Group
4. Conforti, R., de Leoni, M., La Rosa, M., van der Aalst, W.M., ter Hofstede, A.H.: A recommendation system for predicting risks across multiple business process instances. Decis. Support Syst. **69**, 1–19 (2015)
5. Damaggio, E., Hull, R., Vaculin, R.: On the equivalence of incremental and fixpoint semantics for business artifacts with guard-stage-milestone lifecycles. Inf. Syst. **38**(4), 561–584 (2013)
6. Decision Management Solutions, et al.: Decision Model and Notation (DMN), v1.2 (2019), OMG Document Number formal/19-01-05, Object Management Group
7. Di Ciccio, C., Marrella, A., Russo, A.: Knowledge-intensive processes: characteristics, requirements and analysis of contemporary approaches. J. Data Semant. **4**(1), 29–57 (2015)
8. Eshuis, R., Firat, M.: Modeling uncertainty in declarative artifact-centric process models. In: Daniel, F., Sheng, Q.Z., Motahari, H. (eds.) BPM 2018. LNBIP, vol. 342, pp. 281–293. Springer, Cham (2019). https://doi.org/10.1007/978-3-030-11641-5_22
9. Eshuis, R., Hull, R., Yi, M.: Reasoning about property preservation in adaptive case management. ACM Trans. Internet Technol. **19**(1), 12:1–12:21 (2019)
10. Ghattas, J., Soffer, P., Peleg, M.: Improving business process decision making based on past experience. Decis. Support Syst. **59**, 93–107 (2014)
11. Grossmann, W., Rinderle-Ma, S.: Fundamentals of Business Intelligence. Data-Centric Systems and Applications. Springer, Heidelberg (2015). https://doi.org/10.1007/978-3-662-46531-8
12. Hauder, M., Pigat, S., Matthes, F.: Research challenges in adaptive case management: a literature review. In: 2014 IEEE EDOC, pp. 98–107. IEEE (2014)
13. Hull, R., et al.: Introducing the guard-stage-milestone approach for specifying business entity lifecycles. In: Bravetti, M., Bultan, T. (eds.) WS-FM 2010. LNCS, vol. 6551, pp. 1–24. Springer, Heidelberg (2011). https://doi.org/10.1007/978-3-642-19589-1_1
14. Lakshmanan, G.T., Shamsi, D., Doganata, Y.N., Unuvar, M., Khalaf, R.: A Markov prediction model for data-driven semi-structured business processes. Knowl. Inf. Syst. **42**(1), 97–126 (2015)
15. Lim, C., Bearden, J.N., Smith, J.C.: Sequential search with multiattribute options. Decis. Anal. **3**(1), 3–15 (2006)
16. Mertens, S., Gailly, F., Poels, G.: Towards a decision-aware declarative process modeling language for knowledge-intensive processes. Expert Syst. Appl. **87**, 316–334 (2017)
17. Nigam, A., Caswell, N.S.: Business artifacts: an approach to operational specification. IBM Syst. J. **42**(3), 428–445 (2003)
18. Petrusel, R.: Using Markov decision process for recommendations based on aggregated decision data models. In: Abramowicz, W. (ed.) BIS 2013. LNBIP, vol. 157, pp. 125–137. Springer, Heidelberg (2013). https://doi.org/10.1007/978-3-642-38366-3_11
19. Provost, F., Fawcett, T.: Data Science for Business: What You Need to Know About Data Mining and Data-analytic Thinking. O'Reilly, Newton (2013)
20. Puterman, M.L.: Markov Decision Processes: Discrete Stochastic Dynamic Programming. Wiley, Hoboken (2014)
21. Rozinat, A., Wynn, M.T., van der Aalst, W.M., ter Hofstede, A.H., Fidge, C.J.: Workflow simulation for operational decision support. Data Knowl. Eng. **68**(9), 834–850 (2009)

22. Schonenberg, H., Weber, B., van Dongen, B., van der Aalst, W.: Supporting flexible processes through recommendations based on history. In: Dumas, M., Reichert, M., Shan, M.-C. (eds.) BPM 2008. LNCS, vol. 5240, pp. 51–66. Springer, Heidelberg (2008). https://doi.org/10.1007/978-3-540-85758-7_7

23. Seber, G.A., Lee, A.J.: Linear Regression Analysis, vol. 329. Wiley, Hoboken (2012)

24. Vaculin, R., Hull, R., Heath, T., Cochran, C., Nigam, A., Sukaviriya, P.: Declarative business artifact centric modeling of decision and knowledge intensive business processes. In: 2011 IEEE EDOC, pp. 151–160. IEEE (2011)

25. Vanderfeesten, I., Reijers, H.A., Van der Aalst, W.M.: Product-based workflow support. Inf. Syst. 36(2), 517–535 (2011)

26. Voorberg, S., Eshuis, R., van Jaarsveld, W., van Houtum, G.J.: Appendix to: Decision Support for Declarative Artifact-Centric Processes (2019). http://is.ieis.tue.nl/staff/heshuis/dsdacp.pdf

Execution

Optimized Resource Allocations
in Business Process Models

Sven Ihde[1]([✉]), Luise Pufahl[1], Min-Bin Lin[2], Asvin Goel[2], and Mathias Weske[1]

[1] Hasso Plattner Institute, University of Potsdam, 14482 Potsdam, Germany
{sven.ihde,luise.pufahl,mathias.weske}@hpi.de
[2] Kühne Logistics University, Hamburg, Germany
{min-bin.lin,asvin.goel}@the-klu.org

Abstract. The allocation of resources to process activities can have a huge influence on overall performance, in particular, if resources are costly and limited in their availability. Rule-based allocations can lead to unnecessarily low resource utilization rates, high costs, and large delays. In this paper, we present a framework allowing for optimized resource allocations by extending a traditional Business Process Management System by a new component that we call the *Resource Manager*. Our framework allows a process designer to specify resource requirements which are used by the Resource Manager to decide on allocations of resources to process activities. We describe the functionality of the Resource Manager, its interaction with the process engine, and the data needed. The framework is implemented by extending an open-source process modeler and engine, and applied to a use case concerning the last mile delivery.

Keywords: Resource allocation · Resource optimization ·
Business process modeling · Business Process Management System

1 Introduction

Business processes are indispensable for organizations to achieve their goals for providing goods and services. As many resources are cost-intensive and limited [3], process performance is highly dependent on an efficient allocation of resources to tasks. Inefficient resource allocation can lead to low resource utilization, high costs, large delays, and low quality [7]. Resource allocation ensures that each activity of a certain process instance (i.e., a task) is executed at the right time and with the right resources, thereby balancing the demand of process execution with the availability of resources [21].

Business Process Management Systems (BPMSs) coordinate process activities, resources, and data given in a process model and allow process automation [13,38]. Existing BPMSs provide two basic allocation mechanisms: push and pull [34]. Using the push mechanism, tasks are pushed to qualified resources based on pre-defined rules (e.g., the FIFO mechanism) that are specified at design time [8,33]. A summary of those are given by the so-called resource allocation patterns [34]. Using the pull mechanism, tasks are provided in a task list to

© Springer Nature Switzerland AG 2019
T. Hildebrandt et al. (Eds.): BPM 2019, LNBIP 360, pp. 55–71, 2019.
https://doi.org/10.1007/978-3-030-26643-1_4

resources (usually human actors), such that they can prioritize and select based on their knowledge. These two approaches might be useful for several use cases, but are too limited for the domains such as logistics, healthcare, or manufacturing, with complex process structures and costly resources requiring advanced resource allocation mechanisms [18,29]. In the SMile project[1], which motivates this research, cost-efficient vehicle routes must be generated for delivering parcels with promised delivery time windows.

In the business process management (BPM) domain, the resource perspective was intensively studied [5]. Especially the definition of resource requirements is supported (see [3,9]). Different approaches were developed to improve resource allocations by generating an optimized resource allocation plan for a process instance [18], by mining resource allocation rules [24], or by dynamically allocating resources based on historic logs [21]. However, most of them are stand-alone solutions for specific problems and not necessarily integrated into the process configuration and execution. In the community, the need for a resource-aware BPMS being capable of smart resource allocation has been identified [8,29].

In this paper, we present a new component called *Resource Manager* that extends the traditional BPMSs and is responsible for the allocation of resources to tasks. We describe how the Resource Manager can be integrated into a traditional BPMS and which information must be provided by the process modeler. Our proposed architecture allows to provide use cases with specific constraints and performance measures that can be applied to optimize resource allocations. The Resource Manager can rely either on a human decision maker who manually allocates resources to tasks, or on an optimization algorithm for the automatic allocation of resources. To demonstrate the practicality of this research the Resource Manager is implemented to a real-world logistics process.

In the reminder, existing works about the resource perspective of business processes are discussed in Sect. 2. After describing a motivating example Sect. 3, some background on resource-constrained scheduling problems is given in Sect. 4 and requirements are discussed in Sect. 5. Section 6 describes the extended architecture of a resource-aware BPMS and shows how resource allocations can be optimized by a resource manager. A prototypical of the extended architecture and its application to a real-world use case are presented in Sect. 7. Limitations and future work are discussed in Sect. 8.

2   Related Work

In addition to the control-flow and data-flow perspectives of business processes, the resource perspective is essential for the successful execution of business processes [13]. Cabanillas [8] distinguishes among three key operations of resource management: (1) resource assignment (i.e., definition of resource requirements for process activities at design time), (2) resource allocation (i.e., designation of concrete resources to a specific task during runtime), and (3) resource analysis (i.e., evaluation of process execution with the focus on resources).

---

[1] http://smile-project.de.

Recent works have focused on resource meta-models for a detailed specification of resource characteristics to support resource assignment, such as the organizational meta-model [34], resource meta-model [27], or resource classification role-model [29]. A taxonomy of the different approaches is given by Arias et al. [3]. Advanced textual and graphical resource assignment languages were developed in [9] that go beyond simple role-assignment being available in the standard process modeling notation BPMN [28].

Russel et al. [34] introduced 43 workflow resource patterns; a comprehensive study on use and representation of (mostly) human resources in the existing BPMSs. It also included allocations patterns (i.e., role-based, history-based, etc.), which are greedy solutions that may become sub-optimal, especially as resources are not independent from each other [18]. Batch activities developed by [32] efficiently combine the execution of certain activities from several process instances to optimize resource utilization. However, these approaches only support a fixed set of allocation algorithms and only one type of resource. Considering complex resource constraints, Senkul et al. [36] and Havur et al. [18] use logic programming to identify optimal resource allocation plans for deterministic process instances.

Based on historical execution logs, it is possible to generate allocation rules using machine learning techniques [24]. Liu et al. [25] gives resource allocation recommendations to a human assigner during execution based on historical logs with a supervised machine learning algorithm that does not consider the availability or workloads of resources. This work is extended by Arias et al. [4] considering the current workloads of resources, and by Zaoh et al. [39] applying a heuristic to handle several process instances running in parallel. Huang et al. [21] uses reinforcement learning for dynamic resource allocation and considers the influence of resource allocations on the process environment.

In summary, various approaches for resource allocations have been proposed already. However, they are mainly based on rule-based approaches and efficient resource allocation is derived from a business process perspective. From the literature on optimization we know that rule-based approaches usually cannot guarantee an efficient utilization of resources. Additionally, as organizations have multiple processes working on the same limited set of resources, we propose to plan resource allocation form a resource point of view. The Operation Research community has a long history of developing approaches for optimizing resource allocations that way. In order to leverage on these achievements, we propose a framework which integrates sophisticated solution techniques developed in the Operations Research community into traditional BPMSs.

3   Motivating Example

Online shopping has led to an increase in parcel deliveries. The delivery of parcels in the last mile (from the final hub to the recipients) imposes various challenges on logistics service providers in particular due to low profit margins and a high risk of not being able to deliver parcels if recipients are not at home [35]. The

SMile projects aims to innovate the last mile logistics by ensuring a delivery at the first try in a so-called *micro-depot*. In a micro-depot a parcel can be delivered by small and local delivery services to a recipient at a desired time frame. This section presents this last mile delivery process shown in Fig. 1.

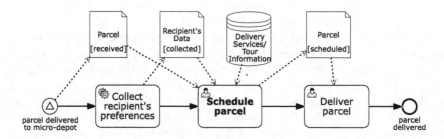

Fig. 1. Process model of parcel deliveries with promised delivery time windows.

When a parcel is delivered to a micro-depot, the recipient's preferences are collected. Based on the preferences, the delivery is scheduled and when the scheduled time is reached, the parcel is delivered to the customer. In order to efficiently utilize the delivery vehicles, the scheduling of individual parcels must be conducted in such a way that cost-efficient routes for vehicles can be found, allowing to deliver parcels at their scheduled times. Thus, the scheduling represented as a simplified user task (cf. Fig. 1) is a complex task, where a suitable vehicle (i.e., resource) and other available parcels in the micro-depot (i.e., other process instances) should be considered. In Fig. 1 it is assumed that this task is done manually by a dispatcher, who has knowledge about available resources and delivery requests. However, the task could also be executed by an engine using appropriate optimization algorithms. The use of such algorithms would not only help in improving process performance by allowing for better delivery routes, it would also allow for full automation of decision making.

This example shows that business processes often include process activities for which an optimized scheduling is required considering the currently running process instances and available resources. This can be also observed in other domains where resources must be allocated to process activities subject to the limitations in the way they can be used.

4   Resource-Constrained Scheduling Problems

The scheduling task of our motivating example requires the solution of a special case of a *resource-constrained scheduling problem* (RCSP). RCSPs are problems in which tasks must be scheduled, and where the execution of the tasks depends on one or more resources and their limited availability [17]. The usual goal of RCSPs is to find a schedule for all relevant tasks to minimize the total resource utilization cost. However, other optimization goals like minimizing the time required for executing all tasks or tardiness may also be specified.

For some RCSPs optimal solutions can be found by using exact approaches based on mixed-integer linear programming [22] and branch-and-bound algorithms [12]. However, as these problems are usually $\mathcal{NP}$-hard [23], they are notoriously difficult to solve, in particular for problems of relevant size. Therefore, RCSPs are usually solved using so-called metaheuristics [30]. While it is possible to search for solutions of RCSPs using rule-based approaches, the solutions are usually of poor quality [6] resulting in unnecessarily high costs, as rule-based approaches can be interpreted as very primitive heuristics. Sophisticated heuristics systematically exploring the search-space and escaping local optima of poor quality can often find near-optimal solutions in reasonable amount of time [14].

In the motivating example, the limited resources are vehicles. Vehicles can conduct multiple parcel deliveries in one route, however, they are limited concerning the total volume and weight of the parcels. Furthermore, different parcels have different destinations and different required times of the delivery. This heterogeneity of the tasks results in the need to plan the order of deliveries and the allocation of vehicles to delivery requests. Poorly planned schedules result in unnecessarily long (and costly) travel times between delivery locations and an unnecessarily high number of vehicles required. An overview of optimization approaches that can be used for similar problems related to the transportation of goods can be found in [37].

While the remainder of this paper focuses on our motivating example, the framework presented in this paper is not restricted to this particular example. For other application scenarios, similar RCSPs may have to be considered, for example, in manufacturing or healthcare. In the following we provide brief examples of typical tasks and resource-requirements in such domains to illustrate the variations of application scenarios that can benefit from our framework.

Manufacturing. Resources in the manufacturing industry are broadly defined as production units, e.g., machines, human operators, material handling vehicles, or storage buffers, which are limited. Multiple tasks are required to be performed by resources to create desired goods. According to types of manufacturing systems, operational processes can be altered [1]. For example, a job shop, which produces goods in high variety but low volume, involves diverse tasks that consist of several operations which have to be processed on different resources, e.g., tool and die making. In contrast, in a flow shop, tasks have to be processed in a consistent order on all resources, e.g., automobile assembly line. Additionally, batch production can occur in any manufacturing system, where components or goods proceed to various stages, in groups of a certain size, to different resources. Thus, resource allocation here is highly related to the design of manufacturing systems. Some common objectives are throughput, makespan, lead time, and resource utilization considering line balancing.

Healthcare. In healthcare services, resources typically refer to doctors, nurses, equipment, and rooms. As available healthcare resources are considerably limited in comparison to patient demands, effectively allocating resources is crucial in enhancing service quality for patients. Moreover, doctors and nurses are restricted in their sub-specializations, for example: surgeon or orthopaedist. The patient groups are likewise bound in their needs to a specialized professional and are thus required to utilize a specific resource [11,26]. Medical staff are generally shared between several patients and are often blocked by who can be served when, and the same is true for equipment and rooms. Hence, to coordinate resource availability with the right resource skills to patient demands, multi-skill resources should be considered in the healthcare resource allocation [2]. General objectives in this class of problems are patient waiting time, makespan, resource utilization, and the choice of doctor preferences.

Given the heterogeneity of RCSPs, no single model can be defined that is suitable for all possible application scenarios. However, all problems have in common that it usually benefits from using sophisticated optimization algorithms exploiting knowledge about the entirety of all tasks and all resources compared to simply assigning tasks to resources one by one.

5   Requirements

Usually multiple process instances simultaneously and concurrently require limited resources. Although greedy rules as provided by the workflow patterns [34] or mined from historical data could be used to allocate resources, such approaches often lead to unnecessary inefficiencies as discussed in the previous section. In order to optimize resource allocations within business processes, a BPMS should provide the possibility to configure and solve a RCSP capturing the characteristics of the specific use case. For this we can identify multiple requirements that should be fulfilled by a BPMS:

R1 *Knowledge on resources and their availability.* To enable an optimized resource allocation, the BPMS needs to have knowledge of the resource characteristics as well as their current or future availability [29]. Therefore, static and dynamic information concerning the resources must be captured.

R2 *Resource requirements.* The BPMS must allow a process designer to specify when which resource is required in which quantity. For more complex application scenarios multiple resources may be required simultaneously. Thus, it should be able to define the information in a process model.

R3 *Constraints and performance measures.* The BPMS must allow a designer to specify the constraints on the utilization of resources and the impact of resource utilization on the overall performance. The constraints and performance measures define the problem of allocating resources.

R4 *Allocation service.* While the problem definition specifies the constraints and goals, it does not describe how the allocation is conducted. As the allocation of resources can be conducted in many different ways, the BPMS must allow the specification and implementation of one or several alternative allocation services that can be selected by the designer.

Given that the problem of allocating resources to tasks is a complex problem that can be tackled in many different ways (e.g. manually or by the use of optimization algorithms), we can observe that there is a need for separation of concerns between the process and resource allocation aspects.

6   Optimized Resource Allocation in Business Processes

In this section we propose to extend the traditional BPMS with additional components for resource management that completely encapsulate the knowledge of resources and their optimized allocation. The extension to the resource components and their interaction with traditional components is presented on the architectural level in Sect. 6.1. In Sect. 6.2 we show how the current process modeling needs to be extended in order to specify the requirements on the optimized resource allocation for process execution. Finally, in Sect. 6.3, the detailed functionality while executing a *resource-allocation* activity is explained.

6.1   Architecture of Resource-Aware BPMS

In Fig. 2, the extended BPMS architecture for realizing optimized resource allocation is presented. It consists of the traditional components of a BPMS and the extension for the resource components. First, we shortly repeat the structure of a traditional BPMS before we introduce the resource-aware extension with the *Resource Modeling* and the *Resource Manager*.

Traditional BPMS. In general, a BPMS consists of four main components: *Process Modeling*, *Process Engine*, *Environment*, and *Applications* [13,38]. The *Process Modeling* component supports the *Process Designers* to model their business processes with all the relevant technical information for an automatic process execution, such as user interfaces, services to-be called, or resource assignment information. This is stored in a *Process Model Repository* and can be also updated. The *Process Engine* is the main component for guiding the process execution, which can access the repository. An instance of a process can be manually started or automatically triggered in the process engine. Then, the process engine initiates the process activities in their prescribed sequence. Business processes involve different *Process Participants* in order to execute/check/document certain process steps. The participants are connected mostly via a User Interface that shows them the corresponding tasks to be done summarized as *Environment*. In case of service activities in a process, other services might be called grouped together as *Applications* representing of all kinds of services.

Resource-Aware Extension. The resource-aware extension of a BPMS consists of two components: *Resource Modeling* and *Resource Manager*.

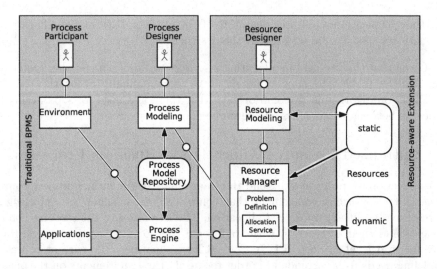

Fig. 2. Architecture of a resource-aware BPMS for smart resource allocation (The diagram is a FMC block diagram (http://www.fmc-modeling.org) representing active software components as rectangles and passive data storage as ellipses).

Resource Modeling. The purpose of the *Resource Modeling* component is to define and store all relevant information concerning resources and the constraints on the resource utilization. Similar to process modeling, an expert is assumed, the *Resource Designer*, who is responsible for prescribing the information and constraints on resources, referred to as static resource data. The static resource data includes, for example, the cost of a resource per use/time-frame, its maximum capacity, type of resource (e.g., reusable, consumptive, or producible). Different organizational resource models were proposed [27,29] which could be used as starting point for a structure of the static resource data. However, use case-specific resource models are required in many cases.

Resource Manager. The purpose of the *Resource Manager* component is to decide on the allocation of resources to process tasks. A resource manager can handle multiple problem definitions based on the constraints provided during resource modeling. For each problem definition, the resource manager provides an interface in which the requested input and expected output for a resource allocation request is defined. During process modeling, dedicated *resource-allocation* activities (further discussed in the next subsection) can be used to request a resource allocation for a selected problem definition. For each problem definition one or several *Allocation Services* can be implemented. An allocation service can be a simple rule-based algorithm, a heuristic, a sophisticated optimization algorithm, a mined allocation or even an interactive component combining algorithms with expert knowledge via the user interface. At process design time, the allocation service can be selected for a resource-allocation activity.

Whenever a resource-allocation activity is executed by the process engine, it maps the respective data objects of the activity to the required inputs and expected outputs of the problem definition. The data is assumed to contain all relevant information allowing the resource manager to update information about current and future availability of resources, i.e., dynamic resource data.

6.2 Configuration

In the following, we define dedicated activities for resource allocations and their configuration. Assume we are given a set of business process models P modelled by the process designer and a set of problem definitions R provided by a resource manager. For each process model $p \in P$ let A_p denote the activities of the process model. Then, the process designer can specify a subset $A_p^* \subset A_p$ representing dedicated activities for requesting the allocation of resources. We refer to these activities as *resource-allocation activities*. For each of these resource-allocation activities $a \in A_p^*$ the process designer must select a problem definition $r \in R$.

For each problem definition r, the required input I_r and expected output O_r is defined, as a set of data attributes $i = (k_i, t_i)$ where k_i is a key that can be used to identify a data item and t_i specifies the data type. These data types can be elementary types or arbitrary complex data types. Section 7.2 shows the data definition with non-elementary data types for our motivating example.

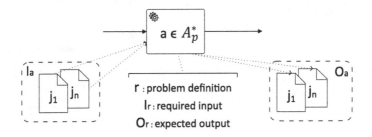

Fig. 3. *Resource-allocation* activity shown as BPMN activity.

The process designer ensures that the data input and output of a resource-allocation activity a is aligned with the input and output of the problem definition r selected for the activity a. The process designer similarly defines the allowed data input I_a for the activity a as a set of process variables $j = (k_j, t_j)$. The proposed resource-aware BPMS checks that for each data attribute $i \in I_r$, a process variable $j \in I_a$ exists for which t_j can be mapped to t_i. Furthermore, the resource-aware BPMS checks that for each data attribute $i \in O_r$, a process variable $j \in O_a$ exists for which t_i can be mapped to t_j. Thus, any output of the resource manager can be mapped to data output of the resource-allocation activity. Figure 3 illustrates the data in- and output.

During process execution, the process engine generates a data input for each resource-allocation activity a that can be represented by a set of process instance variables $j = (k_j, v_j)$ where k_j is the key of the process variable and v_j is the value which must be of the correct type t_j according to the process variable for key k_j. This data input is mapped to the expected input of the problem definition which generates an output that is then mapped to the output of the resource-allocation activity.

The advantage of this proposed configuration is, that it provides a high degree of flexibility. The process designer may, for example, specify all the data that is available and potentially relevant for the resource allocation. A simple allocation service, that may be initially deployed, may not require all the data and may work on a limited subset thereof. If, at a later stage, a more sophisticated allocation service with higher data requirements shall be deployed, the process model does not need to be changed.

6.3 Functionality

The execution of the resource-allocation activity at runtime and the interaction between the process engine and resource manager is visualized as a UML sequence diagram in Fig. 4.

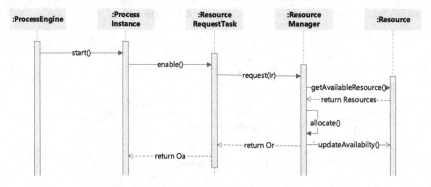

Fig. 4. Interaction between the Process Engine and Resource Manager for executing a *resource-allocation* activity shown as UML sequence diagram.

The `ProcessEngine` instantiates a process model to start a new `Process Instance`. During the execution of the process instance, activities are enabled and executed based on the prescribed order in the process model. In the case of a `Resource AllocationActivity`, the selected interface of the `ResourceManager` is called with the input data I_r for the respective optimization problem. The `ProcessInstance` is now in a waiting state until the request is answered by the `ResourceManager`. After receiving a request the `ResourceManager` collects the information on the resource(s), which can be used to fulfill the request, and checks their availability. In the next step, the `ResourceManager` uses the allocation service for the selected problem definition in order to find a suitable resource allocation. The `ResourceManager` collects multiple requests and conducts the resource allocation for these requests together. After running the resource allocation, it returns the resource(s) O_r to the requesting *resource-allocation* activity. The resource(s) are now reserved for use in the process instance. With the information on the planned usage, the `ResourceManager` updates the availability of the resource. The entry is a timeframe, when the resource is not available in future and cannot be used for any other allocations in this specific timeframe.

Currently, we assume that the usage of the resources is executed as planned by the resource manager. However, for treating delays and exceptions the resource manager needs to be informed about the actual start and the end of usage of resources by process instances. It could be realized, for example, by having events published by the process engine about the start and end of process activities as described in [19] to which the resource manager can subscribe. Then the resource manager can compare the actual start and end to the planned allocation information. If there are any variations to the plan, the resource manager might need to adapt previous plannings. This exception handling will not be the focus of the current work.

7 Evaluation

This section presents an evaluation of the resource-aware BPMS extension in a two-fold manner: (1) a prototypical implementation is presented in Sect. 7.1, and (2) an application of the approach to a logistics use case is illustrated in Sect. 7.2.

7.1 Prototypical Implementation

As shown in Fig. 5, we used for our prototypical implementation *Chimera*[2] [20] – a process engine for flexible, knowledge-intensive processes. It also provides a process modeler – *Gryphon*, which is based on the open-source BPMN modeler *bpmn.io*. As soon as the process model is designed and configured in *Gryphon*, it can be stored in the *Process Models* repository and can be deployed to *Chimera* engine as depicted in Fig. 5. As these two components represent a traditional BPMS, we used it as a base and adapted it to fit our architecture proposal in Fig. 2. For this we implemented a simplified Resource Manager, the *Sphinx*[3] that is connected to *Chimera* and *Gryphon* via REST interfaces.

The interface between *Sphinx* and *Gryphon* is used during design time and provides the information a process designer needs for configuring the task, as shown in Fig. 6. That is why we extended *Gryphon*[4] to allow the configuration of resource-allocation activities. It allows to specify the Resource Manager, a problem definition, and the allocation service that is used (see requirements R2

---

[2] https://bptlab.github.io/chimera/.

[3] https://github.com/bptlab/smile/tree/master/sphinx.

[4] https://github.com/bptlab/gryphon/tree/resource/add-resource-type.

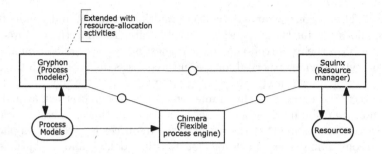

Fig. 5. Architecture of prototypical implementation

and R4). Based on this selection the corresponding input and output sets are shown, that have to be mapped by the designer.

If *Chimera* reaches a resource-allocation activity during the process execution, then it calls the referenced Resource Manager with the defined input. Based on the steps described by Fig. 4 the Resource Manager will access its resource information to find the best possible allocation, as shown in the next section.

7.2   Application for Our Motivating Example

Problem Definition and Allocation Algorithm. For an efficient last-mile delivery, delivery routes must be found which minimize total distance travelled. In our motivating example, the deliveries must be conducted between an earliest and a latest given delivery time. The optimization problem for finding efficient delivery routes is a so-called *vehicle routing problem with time windows (VRPTW)* [37]. A rule-based solution would, for example, allocate the next available vehicle to a request. As parcels can arrive at the micro-depot at any time, not all delivery requests are known to the Resource Manager when a vehicle is requested, thus resulting in the rule-based approach leading to an overall poor solution. Instead routes have to be created given all the information known to date and revised when new information about request becomes available, which is possible with a heuristic. This variant of the vehicle routing problem is known as a *dynamic vehicle routing problem (DVRP)* [31].

In our implementation we used an *insertion method* [10] as the allocation service. An insertion method successively selects unscheduled requests and adds them to the route of a vehicle. For each request the method determines all feasible insertion possibilities. That is, for each vehicle route and each request already in the route, the algorithm checks whether the unscheduled request can be added to the route before or after the already scheduled request. If the request can be feasibly inserted, the method inserts the request at the position in the route which has the lowest incremental costs. If no feasible insertion possibility is found, a new route is created and the request is added into the new route.

To handle the dynamics of parcels arriving at the micro-depot, the deterministic VRP is iteratively solved with a rolling-horizon procedure [15] in which the optimization period moves forward at each iteration and adds each time period to the total planning time horizon. The rolling-horizon procedure allows the schedule for the previous planning time period, which is actually implemented, to be influenced by the future in the form of the next planning time period. The insertion method is a straightforward strategy that can easily adapt complex constraints and generate a feasible solution within a short computation time.

In this paper, we use the insertion method due to its simplicity and ease of implementation, as our goal is to provide a proof-of-concept demonstrating how an algorithm is used for resource allocations. However, it is easy to replace the allocation service by more powerful algorithms for dynamic vehicle routing [31] or by manual or even interactive dispatching systems [16]. From the literature on vehicle routing problems it is known that the performance of delivery processes can be significantly improved by state-of-the-art planning approaches.

Configuration of the Resource Allocation Task for Parcel Delivery. The process of our motivating example shown in Fig. 1 can be easily adapted to leverage the functionality of our resource manager and to fully automate the scheduling of parcels.

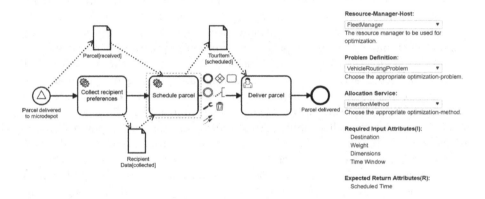

Fig. 6. Configuration in Gryphon

As shown in Fig. 6 we replace the user task for scheduling the parcel by a service task. This service task is a resource-allocation task for which we implemented a configuration sidebar allowing the selection of a *Resource Manager Host*, a *Problem Definition*, and an *Allocation Service*. We implemented a resource manager named *Fleet Manager* that is responsible for the allocation of vehicles in the fleet. The *Fleet Manager* provides a list of problem definitions and allocation services. From this list the process designer can choose the *Vehicle Routing Problem* as problem definition and the *Insertion Method* as allocation service. The chosen resource managers defines the *Required Input*, and *Expected Return* which are shown to the process designer in the sidebar.

The input definition provided by the resource manager includes data definition items *("Destination"*,(float,float)) for the geographic coordinates of the destination of the parcel, *("Weight"*,float) for the weight of the parcel, *("Dimensions"*, (float,float,float)) for the width, height, and length of the parcel, and *("TimeWindow"*, (time,time)) for the earliest and latest delivery times. The process designer has to make sure that for each process instance (i.e., each parcel to be delivered) the respective data items are generated during process execution and provided to the resource-allocation task by data objects. The output definition provided by the resource manager includes the data definition item *("Scheduled Time"*, time) representing the scheduled delivery time. The process designer can utilize the respective output by passing a data object from the resource-allocation activity to subsequent activities for further utilization. In our example, the data item for the scheduled delivery time is included into a data object that is used by the activity *Deliver parcel*.

8   Conclusion

In this paper, we provide a framework to integrate optimized resource allocations in business processes by extending the traditional BPMS architecture through a component called the Resource Manager. The Resource Manager is responsible for maintaining all relevant information concerning the availability of resources and for allocating resources to a process instance. The process designer can specify resource requirements within the business process model through dedicated resource-allocation activities. By implementing and running our approach for a use case concerning time-constrained parcel deliveries, we demonstrate that the proposed architecture can be used to integrate complex allocation services in business processes. This enables process automation by using powerful operations research techniques. An advantage of our approach is, that it allows a process designer to focus on specifying the requirements on resources, whereas an expert in optimization can focus on how resources are best allocated to process instances. A change of allocation services (e.g., an optimization algorithm, a simple heuristic, machine learning algorithms) is easily possible without adaptation of the respective process model. In the future, historical execution data of resource allocations could also be used for learning and selecting an appropriate allocation service. Currently, we assume that the process execution always follows the allocation plan. This could be extended by considering real-time execution data to adapt the allocation plans automatically.

Acknowledgements. The research leading to these results has been partly funded by the BMWi under grant agreement 01MD18012C, Project SMile http://smile-project.de.

References

1. Abedinnia, H., Glock, C.H., Grosse, E.H., Schneider, M.: Machine scheduling problems in production: a tertiary study. Comput. Ind. Eng. **111**, 403–416 (2017)
2. Ağralı, S., Taşkın, Z.C., Ünal, A.T.: Employee scheduling in service industries with flexible employee availability and demand. Omega **66**, 159–169 (2017)
3. Arias, M., Munoz-Gama, J., Sepúlveda, M.: Towards a taxonomy of human resource allocation criteria. In: Teniente, E., Weidlich, M. (eds.) BPM 2017, vol. 308, pp. 475–483. Springer, Cham (2018). https://doi.org/10.1007/978-3-319-74030-0_37
4. Arias, M., Rojas, E., Munoz-Gama, J., Sepúlveda, M.: A framework for recommending resource allocation based on process mining. In: Reichert, M., Reijers, H.A. (eds.) BPM 2015. LNBIP, vol. 256, pp. 458–470. Springer, Cham (2016). https://doi.org/10.1007/978-3-319-42887-1_37
5. Arias, M., Saavedra, R., Marques, M.R., Munoz-Gama, J., Sepúlveda, M.: Human resource allocation in business process management and process mining: a systematic mapping study. Manag. Decis. **56**(2), 376–405 (2018)
6. Bang-Jensen, J., Gutin, G., Yeo, A.: When the greedy algorithm fails. Discrete Optim. **1**(2), 121–127 (2004)
7. Bellaaj Elloumi, F., Sellami, M., Bhiri, S.: Avoiding resource misallocations in business processes. Concurrency Comput.: Practice Exp. e4888 (0000). https://doi.org/10.1002/cpe.4888
8. Cabanillas, C.: Process-and resource-aware information systems. In: 2016 IEEE 20th International EDOC, pp. 1–10. IEEE (2016)
9. Cabanillas, C., Resinas, M., Ruiz-Cortés, A.: RAL: a high-level user-oriented resource assignment language for business processes. In: Daniel, F., Barkaoui, K., Dustdar, S. (eds.) BPM 2011. LNBIP, vol. 99, pp. 50–61. Springer, Heidelberg (2012). https://doi.org/10.1007/978-3-642-28108-2_5
10. Campbell, A.M., Savelsbergh, M.: Efficient insertion heuristics for vehicle routing and scheduling problems. Transp. Sci. **38**(3), 369–378 (2004)
11. Cardoen, B., Demeulemeester, E., Beliën, J.: Operating room planning and scheduling: a literature review. Eur. J. Oper. Res. **201**(3), 921–932 (2010)
12. Coelho, J., Vanhoucke, M.: An exact composite lower bound strategy for the resource-constrained project scheduling problem. Comput. Oper. Res. **93**, 135–150 (2018)
13. Dumas, M., Rosa, M.L., Mendling, J., Reijers, H.A.: Fundamentals of Business Process Management, 2nd edn. Springer, Heidelberg (2018). https://doi.org/10.1007/978-3-662-56509-4
14. Gendreau, M., Potvin, J.Y., et al.: Handbook of Metaheuristics, vol. 2. Springer, Heidelberg (2010). https://doi.org/10.1007/978-1-4419-1665-5
15. Ghiani, G., Guerriero, F., Laporte, G., Musmanno, R.: Real-time vehicle routing: solution concepts, algorithms and parallel computing strategies. Eur. J. Oper. Res. **151**(1), 1–11 (2003)
16. Goel, A.: Fleet Telematics - Real-Time Management and Planning of Commercial Vehicle Operations. Operations Research/Computer Science Interfaces, vol. 40. Springer, Heidelberg (2007). https://doi.org/10.1007/978-0-387-75105-4
17. Hartmann, S., Briskorn, D.: A survey of variants and extensions of the resource-constrained project scheduling problem. Eur. J. Oper. Res. **207**(1), 1–14 (2010)

18. Havur, G., Cabanillas, C., Mendling, J., Polleres, A.: Resource allocation with dependencies in business process management systems. In: La Rosa, M., Loos, P., Pastor, O. (eds.) BPM 2016. LNBIP, vol. 260, pp. 3–19. Springer, Cham (2016). https://doi.org/10.1007/978-3-319-45468-9_1

19. Herzberg, N., Meyer, A., Weske, M.: An event processing platform for business process management. In: 2013 17th IEEE International Enterprise Distributed Object Computing Conference, pp. 107–116. IEEE (2013)

20. Hewelt, M., Weske, M.: A hybrid approach for flexible case modeling and execution. In: La Rosa, M., Loos, P., Pastor, O. (eds.) BPM 2016. LNBIP, vol. 260, pp. 38–54. Springer, Cham (2016). https://doi.org/10.1007/978-3-319-45468-9_3

21. Huang, Z., van der Aalst, W.M., Lu, X., Duan, H.: Reinforcement learning based resource allocation in business process management. Data Knowl. Eng. **70**(1), 127–145 (2011)

22. Kyriakidis, T.S., Kopanos, G.M., Georgiadis, M.C.: MILP formulations for single- and multi-mode resource-constrained project scheduling problems. Comput. Chem. Eng. **36**, 369–385 (2012)

23. Lenstra, J.K., Kan, A.R.: Computational complexity of discrete optimization problems. Ann. Discrete Math. **4**, 121–140 (1979)

24. Liu, T., Cheng, Y., Ni, Z.: Mining event logs to support workflow resource allocation. Knowl.-Based Syst. **35**, 320–331 (2012)

25. Liu, Y., Wang, J., Yang, Y., Sun, J.: A semi-automatic approach for workflow staff assignment. Comput. Ind. **59**(5), 463–476 (2008)

26. May, J.H., Spangler, W.E., Strum, D.P., Vargas, L.G.: The surgical scheduling problem: current research and future opportunities. Prod. Oper. Manag. **20**(3), 392–405 (2011)

27. Oberweis, A.: A meta-model based approach to the description of resources and skills. In: AMCIS 2010 (2010)

28. OMG: Notation BPMN version 2.0. OMG Specification, Object Management Group, pp. 22–31 (2011)

29. Ouyang, C., Wynn, M.T., Fidge, C., ter Hofstede, A.H., Kuhr, J.C.: Modelling complex resource requirements in business process management systems. In: ACIS 2010 Proceedings (2010)

30. Pellerin, R., Perrier, N., Berthaut, F.: A survey of hybrid metaheuristics for the resource-constrained project scheduling problem. Eur. J. Oper. Res. (2019). https://doi.org/10.1016/j.ejor.2019.01.063

31. Pillac, V., Gendreau, M., Guéret, C., Medaglia, A.L.: A review of dynamic vehicle routing problems. Eur. J. Oper. Res. **225**(1), 1–11 (2013)

32. Pufahl, L., Weske, M.: Batch activities in process modeling and execution. In: Basu, S., Pautasso, C., Zhang, L., Fu, X. (eds.) ICSOC 2013. LNCS, vol. 8274, pp. 283–297. Springer, Heidelberg (2013). https://doi.org/10.1007/978-3-642-45005-1_20

33. Reijers, H.A., Jansen-Vullers, M.H., zur Muehlen, M., Appl, W.: Workflow management systems + swarm intelligence = dynamic task assignment for emergency management applications. In: Alonso, G., Dadam, P., Rosemann, M. (eds.) BPM 2007. LNCS, vol. 4714, pp. 125–140. Springer, Heidelberg (2007). https://doi.org/10.1007/978-3-540-75183-0_10

34. Russell, N., van der Aalst, W.M.P., ter Hofstede, A.H.M., Edmond, D.: Workflow resource patterns: identification, representation and tool support. In: Pastor, O., Falcão e Cunha, J. (eds.) CAiSE 2005. LNCS, vol. 3520, pp. 216–232. Springer, Heidelberg (2005). https://doi.org/10.1007/11431855_16

35. Savelsbergh, M., Van Woensel, T.: 50th anniversary invited article-city logistics: challenges and opportunities. Transp. Sci. **50**(2), 579–590 (2016). https://doi.org/10.1287/trsc.2016.0675
36. Senkul, P., Toroslu, I.H.: An architecture for workflow scheduling under resource allocation constraints. Inf. Syst. **30**(5), 399–422 (2005)
37. Toth, P., Vigo, D.: Vehicle Routing: Problems, Methods, and Applications. MOS-SIAM Series on Optimization, no. 18. SIAM, Philadelphia (2014)
38. Weske, M.: Business Process Management - Concepts, Languages, Architectures, 2nd edn. Springer, Heidelberg (2012). https://doi.org/10.1007/978-3-642-28616-2
39. Zhao, W., Liu, H., Dai, W., Ma, J.: An entropy-based clustering ensemble method to support resource allocation in business process management. Knowl. Inf. Syst. **48**(2), 305–330 (2016)

Predicting Critical Behaviors in Business Process Executions: When Evidence Counts

Laura Genga[1](✉), Chiara Di Francescomarino[2], Chiara Ghidini[2], and Nicola Zannone[1]

[1] Eindhoven University of Technology, Eindhoven, The Netherlands
{l.genga,n.zannone}@tue.nl
[2] Fondazione Bruno Kessler, Trento, Italy
{dfmchiara,ghidini}@fbk.eu

Abstract. Organizations need to monitor the execution of their processes to ensure they comply with a set of constraints derived, e.g., by internal managerial choices or by external legal requirements. However, preventive systems that enforce users to adhere to the prescribed behavior are often too rigid for real-world processes, where users might need to deviate to react to unpredictable circumstances. An effective strategy for reducing the risks associated with those deviations is to *predict* whether undesired behaviors will occur in running process executions, thus allowing a process analyst to promptly respond to such violations. In this work, we present a predictive process monitoring technique based on Subjective Logic. Compared to previous work on predictive monitoring, our approach allows to easily customize both the reliability and sensitivity of the predictive system. We evaluate our approach on synthetic data, also comparing it with previous work.

1 Introduction

Today's organizations typically need to monitor their processes to guarantee that they are executed within given boundaries. These boundaries can be set internally, e.g., by process managers, to enhance operational efficiency or can be derived by external legal requirements like the Sarbanes-Oxley Act [1]. As a result, organizations often define execution procedures to ensure that their processes meet these constraints. Deviating from these procedures can expose organizations to abuses and frauds.

However, procedures are often not enforced by design or can be bypassed in order to ensure business continuity [3]. In fact, preventive systems are typically too rigid to deal with real-world, dynamic environments wherein unpredictable circumstances and exceptions often raise. In these settings, a crucial challenge for organizations is to *predict* whether a given (set of) undesired behavior(s) will or will not occur in running process executions, to be able to timely take actions in order to prevent/mitigate potential risks.

© Springer Nature Switzerland AG 2019
T. Hildebrandt et al. (Eds.): BPM 2019, LNBIP 360, pp. 72–90, 2019.
https://doi.org/10.1007/978-3-030-26643-1_5

To address this issue, one can exploit predictive process monitoring [15]. This comprises a family of techniques aimed to predict the "outcome" of running process executions, which in our context corresponds to the occurrence of undesired behaviors. One of the main challenges in predictive monitoring is balancing the reliability and effectiveness of predictions. On the one hand, the predictive system should raise an alarm only when there is "enough" evidence of the forthcoming occurrence of an undesirable behavior. High false-alarms rates, indeed, significantly hamper the usability of such systems, requiring the analyst to waste a large amount of time in verifying the alerts and, thus, leading to a loss of trust in the system. On the other hand, a system that provides late predictions or that, anyway, fails to provide a prediction in most cases, is of little or no use.

Predictive monitoring approaches often allow an analyst to determine the best trade-off between these two forces by acting on a set of metrics that the analyst can customize to her needs. Two commonly-used metrics are the *support* and *confidence* of predictions. The first metric accounts for the amount of history corresponding to the current state of the process execution and it is used to ensure that the prediction is supported by enough evidence. Varying the support threshold impacts the effectiveness of the predictive system; higher the threshold is, more evidence is required to make a prediction. The confidence is instead used to evaluate in which extent the predictive system was able to provide a correct prediction when the given state of execution occurred in the past. This metric impacts both the effectiveness and reliability of the system; requiring high confidence leads to generate predictions only for executions for which high quality predictions had been obtained in the past, thus reducing the number of false positives and false negatives.

However, previous work typically does not allow customizing the *sensitivity* of the prediction; namely, the system returns the outcome that it estimates to be the most likely, without accounting for the "gap" between the probabilities of occurrence/non-occurrence of a behavior. In contrast, a different level of sensitivity may be required in different contexts. For example, in cases where reacting to an alarm is difficult and/or costly, the analyst might be willing to take the risk of waiting to be sure that it is actually necessary to take an action. Ideally, this can be achieved by setting a low sensitivity for the prediction; that is, configuring the system in such a way that it will raise an alarm only when the probability of occurrence is "enough" higher than the probability of non-occurrence. This, however, requires changing the way predictions are usually computed, which is not trivial and even not always possible in existing approaches.

To deal with this challenge, in this work we perform an exploratory study on the application of *Subjective Logic* [12,20] in the context of predictive process monitoring. Subjective Logic is an evidence-based opinion algebra used to evaluate the belief that a given proposition is true or false, explicitly modeling the *uncertainty* in the generation of a prediction. Considering the occurrence of a given behavior as a proposition and past process executions as evidence supporting/contradicting the proposition, Subjective Logic can be used to deter-

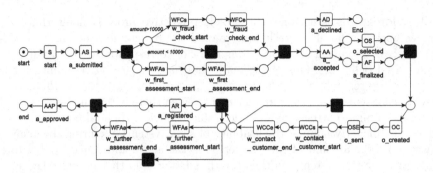

Fig. 1. Loan management process

mine the likelihood that this behavior will or will not occur, or whether there is not enough evidence to make a prediction, thus providing us with a sound and rigorous method to deal with uncertainty.

Elaborating upon Subjective Logic, we introduce a novel prediction approach that allows analysts to customize both the reliability, effectiveness and sensitivity of predictions. We developed a proof-of-concept implementation and tested it over a synthetic dataset to evaluate the validity of the approach and to perform a first assessment of its performance. Results show that our approach is comparable to existing techniques in terms of quality of predictions, while it provides overall better results in terms of effectiveness by being able to make a prediction for a higher number of samples compared to the tested competitor.

The remainder of the paper is organized as follows. Section 2 introduces a running example that is used through the paper. Section 3 describes our approach. Section 4 presents an evaluation of the approach along with a comparison with a well-known predictive monitoring approach. Finally, Sect. 5 discusses related work and draws conclusions.

2   Running Example

Consider, as a running example, a loan management process derived from previous work on the event log of a financial institute made available for the BPI2012 challenge [2,11]. Figure 1 shows the process in Petri net notation. Places are graphically represented by circles and transitions by boxes. Labels below the transitions report the activity names, whereas labels inside the transitions the corresponding acronyms. Black boxes represent *invisible* transitions, i.e. transitions that are not observed by the information systems and are mainly used for routing purposes.

The process starts with the submission of an application. Then, the application passes through a first assessment, aimed to verify whether the applicant meets the requirements. If the requested amount is greater than 10000 euros, the application also goes through a more accurate analysis to detect possible frauds. If the application is not eligible, the process ends; otherwise, the application is accepted. An offer to be sent to the customer is selected and the details of the application are finalized. After the offer has been created and sent to the customer, the latter is contacted to check whether she intends to accept the offer. If this is not the case, the offer is renegotiated and a new offer is sent to the customer. At the end of the negotiation, the agreed application is registered on the system. At this point, further checks can be performed on the application, if needed, before approving it.

Let us assume that two deviating behaviors are allowed in our scenario:

– *Delaying the completion of fraud checking.* Since fraud checking is usually a time-consuming activity, in some cases users can execute other tasks of the process while the checking is not finished yet.
– *Resuming declined applications.* In some cases, a previously rejected application can be resumed, e.g., when the current salary of the customer does not provide enough guarantees for the requested loan amount, but the customer claims that he expects it will be increased. To speed up the process, employees can decide to (temporarily) reject the application, wait that the customer's salary is raised and reuse the previous application, without restarting the process from scratch.

Although these deviating behaviors might be considered acceptable practices, they pave the way to possible abuses. We report below some executions of the process in Fig. 1 in which these behaviors occur:

$$\sigma_1 = \langle S, AS, WFCs, WFAs, WFAe, AA, AF, OS, OC, OSE, WCCs, WCCe, WFCe, AR, AAP \rangle$$
$$\sigma_2 = \langle S, AS, WFAs, WFAe, AD, AA, AF, OS, OC, OSE, WCCs, WCCe, AR, AAP \rangle$$
$$\sigma_3 = \langle S, AS, WFAs, WFAe, AA, AF, OS, OC, OSE, WCCs, WCCe, OC, OSE, WCCs, WCCe, AR, AAP \rangle$$

The first process execution shows that fraud checking has been completed (*WFCe*) only after the offer was sent to the customer (*OSE*). This is clearly undesired. In fact, interrupting an application at this point is costly and likely leads to a loss of the customer's trust. As a result, the employee performing the fraud checking might be more inclined to accept some risks and allow the process to proceed.

Process execution σ_2 shows the management of a declined application (*AD*) that was resumed and then approved (*AAP*) without further assessment. Although the resuming of rejected applications is acceptable, it is highly advisable to perform further assessments on the application before the final approval to verify whether the issues that led to the initial rejection have been solved.

Also compliant behaviors, when misused, might hide possible threats. As an example, σ_3 represents multiple repetitions of the application negotiation with the customer, followed by an approval (*AAP*) without any further assessment. This behavior might be a signal that when the negotiation takes long time, the application is immediately approved when an agreement is reached, without further assessment. An insider might exploit this practice at his own advantage to obtain desirable offers that would not be approved otherwise.

It is worth noting that neglecting the sensitivity of predictions in this kind of scenarios can easily lead to a large number of alerts. For example, if in past process executions the completion of fraud assessment was delayed for slightly more

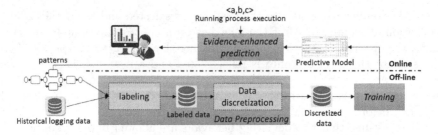

Fig. 2. Approach

applications (with a given range of amounts) than for the ones the check was performed in time, the predictive system may raise an alert for every application with a similar amount. However, it is clearly neither convenient nor feasible performing additional checks for every application only based on the required amount, unless it is *much* more likely that a fraud will occur for applications with given amounts.

3 Approach

The goal of this work is to devise an approach to predict the occurrence of critical behaviors in a running process execution. We assume that the behavior to be monitored is modeled through a set of patterns representing (portions of) process behavior an analyst is interested in.

Our approach, depicted in Fig. 2, follows traditional Machine Learning based approaches to predictive process monitoring and consists of an off-line (training) phase and an on-line (predicting) phase. A characterizing aspect of this work is the employment of Subjective Logic [12] to assess the quality of predictions. Subjective Logic is an opinion algebra that allows assessing the probability that a given pattern occurs by explicitly accounting for uncertainty based on the amount of available evidence. The following sections detail the steps of our approach.

3.1 Data Preprocessing

Process executions are usually recorded in *event logs*. To build the predictive model, the log should be preprocessed in order to obtain a format suitable for the analysis. Next, we first formally define event logs and then we present the preprocessing steps.

Definition 1 (Event, Event Trace, Event Log). *Let A be the set of process activities and V the set of data attributes. Given an attribute $v \in V$, $U(v)$ denotes the domain of v and $\mathcal{U} = \cup_{v \in V} U(v)$ the union of all attribute domains. An event $e = (a, \varphi_e)$ consists of an activity $a \in A$ and a function φ_e that assigns values to attributes $V_e \subseteq V \colon \varphi_e \in V_e \to \mathcal{U}$ s.t. for all v occurring in φ_e $\varphi_e(v) \in U(v)$.*

The set of events is denoted by $\mathcal{E}$. An event trace $\sigma \in \mathcal{E}^*$ is a sequence of events. An event log $\mathcal{L} \in \mathbb{B}(\mathcal{E}^*)$ is a multiset of event traces.[1]

Given an event $e = (a, \varphi_e)$, we use $act(e)$ to denote the activity label associated to e, i.e. $act(e) = a$. This notation extends to event traces. Given an event trace $\sigma = \langle e_1, \ldots, e_n \rangle \in \mathcal{E}^*$, $act(\sigma)$ denotes the sequence of activities obtained from the projection of the events in σ to their activity label, i.e. $act(\sigma) = \langle act(e_1), \ldots, act(e_n) \rangle$.

To build and train a predictive model, we *label* the event log by indicating which patterns occurred in each trace. The problem of determining whether a pattern occurred in a process execution can be modeled as a *compliance checking* problem [10]. In this work, we model the patterns of interest as data Petri nets since several well-known techniques exist to detect the occurrence of this kind of patterns in a process execution (see, e.g., [3,18]). Note, however, that our approach does not pose any constraint on the choice of the patterns formalism, as well as on the technique employed to detect them.

Figure 3 shows four patterns representing undesired behaviors for the process in Fig. 1: *delayed fraud check, application resuming violation, multiple negotiations* and *old application resuming*. Inspired by [11], we use ω transitions as placeholders to specify that at a given point of the process execution any activity can be executed.

The first three patterns represent the behaviors we already discussed in Sect. 2. The last one is a variant of the application resuming pattern in which the *time* between the rejection and resuming of an application is also constrained. The idea behind this pattern is that resuming too old applications might lead to some risk, since the information initially provided by the applicant on his financial situation might have become outdated.

Once past process executions have been labelled, we apply *data discretization* on the labelled data. To generate accurate predictions, we need to take into account those data attributes that are related with the patterns of interest. This is, however, far from trivial. Especially when dealing with numerical attributes, it is not feasible to consider all values an attribute has assumed/can assume in the trace; therefore, we need to employ an effective strategy to *discretize* the attribute domain in a set of finite intervals.

In order to discretize continuous data, we resort to *supervised discretization*. This approach discretizes continuous variables by taking into account the class values, i.e., selecting discretized intervals that best discriminate between positive and negative classes – in our case between the occurrence and the non-occurrence of a pattern. The approach orders the numerical values of each continuous variable in the training set and selects the split point that produces the highest *information gain*, i.e., the amount of information gained by knowing the value of the attribute, to build the discretized intervals.

However, since the conditions that discriminate between positive and negative classes do not only depend on a single variable, we leverage the supervised

---

[1] $\mathbb{B}(X)$ represents the set of all multisets over X.

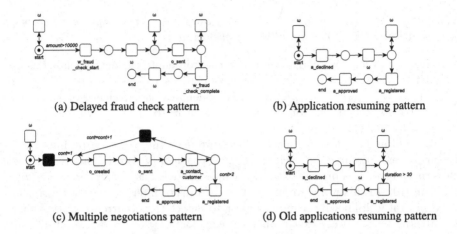

(a) Delayed fraud check pattern

(b) Application resuming pattern

(c) Multiple negotiations pattern

(d) Old applications resuming pattern

Fig. 3. Patterns of undesired behaviors for the loan management process

discretization provided by the decision tree algorithm [17]. This algorithm, in order to build the tree, selects both the variable to be split and the value to be used for the splitting by maximizing the resulting information gain. By inspecting the tree, only the paths root-leaf ending in a leaf with confidence and support enough high, i.e., over user-defined thresholds, are used to extract the discretized intervals. Thus, the retrieved discretized intervals depend on the classes on which they are supposed to discriminate, i.e., on the occurrence of a specific *single-pattern*. However, multilabel classification can be leveraged to retrieve discretized intervals *globally* discriminating on all patterns together. We tested both discretization strategies in our experiments (Sect. 4), to check whether the increased efficiency of the global discretization allows for obtaining predictions as accurate as the ones obtained with different discretization intervals for each pattern (single pattern).

3.2 Training

This step takes as input *(i)* a (preprocessed) event log and *(ii)* the set of patterns the analyst wants to predict, and returns a prediction model representing, for each pattern, the likelihood that the pattern occurs in each "state" of the process. Roughly speaking, the state of a process represents its execution at a given time, i.e. the performed activities along with the value of data attributes. As in [4], we define the state of a process execution as follows:

Definition 2 (State). *Let A be the set of activities, V the set of attributes and U the attributes' domain. A state s for an event trace σ is a pair $(act(\sigma), \varphi_s)$ where φ_s is a function that associates a value to each attribute, i.e. $\varphi_s : V \to U \cup \{\bot\}$ such that for all $v \in V$, $\varphi_s(v) \in U(v) \cup \{\bot\}$ (where $\bot$ indicates undefined). The initial state is denoted $s_I = (\langle\rangle, \varphi_I)$ where φ_I is the initial assignment of values to attributes.*

The initial attribute assignment φ_I represents the value of the attributes in V before the process is executed. For some attributes, a value may not be (initially) defined and, thus, it is considered undefined ($\bot$). The execution of the activities changes the state of the process. We first define the notion of state transition, which is a change of one state to another state due to the effect of an event, and then we extend this definition to event traces.

Definition 3 (State Transition). *Let V be the set of attributes. Given an event $e = (a, \varphi_e)$ and a state $s = (act(\sigma), \varphi_s)$ for an event trace σ, e transforms s into a state $s' = (act(\sigma'), \varphi_{s'})$ such that $\sigma' = \sigma \oplus e$ and for every $v \in V$*

$$\varphi_{s'}(v) = \begin{cases} \varphi_e(v) & \text{if } v \in dom(\varphi_e) \\ \varphi_s(v) & \text{otherwise.} \end{cases} \tag{1}$$

We denote $s \xrightarrow{e} s'$ the state transition given by e.

Intuitively, Eq. 1 states that the occurrence of an event updates the data attributes associated to the event (i.e., $v \in dom(\varphi_e)$) while the other attributes in φ_s remains unchanged.

Definition 4 (Trace Execution). *Given an event trace $\sigma = \langle e_1, ..., e_n \rangle \in \mathcal{E}^*$, σ transforms the initial state s_I into a state s if there exist states $s_0, s_1, \ldots, s_n$ such that*

$$s_I = s_0 \xrightarrow{e_1} s_1 \xrightarrow{e_2} \ldots \xrightarrow{e_n} s_n = s$$

We denote $\mathsf{state}(\sigma)$ the state yielded by an event trace σ.

Missing data attributes introduce uncertainty on the reached state as different states could have been reached. To deal with missing values in an event trace, we adopt the notion of *state subsumption* from [4]. State subsumption is used to determine the possible states of the process that could have been yielded by an event trace.

Definition 5 (State Subsumption). *Given two states $s = (r_s, \varphi_s)$ and $s' = (r_{s'}, \varphi_{s'})$, we say that s subsumes s', denoted $s \succ s'$, if and only if (i) $r_s = r_{s'}$ and (ii) for all $v \in V$ s.t. $\varphi_s(v) \neq \bot$, $\varphi_{s'}(v) = \varphi_s(v)$.*

The table on the right shows the prediction model obtained from (a portion) of σ_1 after attributes *amount* and *duration* have been discretized in the preprocessing step. States consist of the prefixes of the trace along with the attribute values (or discretization intervals) obtained after the partial process execution corresponding to the prefix. The occurrence of an event may or may not change the value of an attribute. For

State				Pattern	#
Executed activities	Data attributes				
	amount	*duration*	...		
$\langle \rangle$	A_1	D_1	...	π_1	5
				π_2	0
$\langle S \rangle$	A_1	D_2	...	π_1	5
				π_2	0
$\langle S, AS \rangle$	A_1	D_3	...	π_1	5
				π_2	0
$\langle S, AS, WFAs \rangle$	A_1	D_4	...	π_1	3
				π_2	0
...	...	...	...	...	...

instance, the *amount* does not change during a process execution, whereas the *duration* of the process execution is updated after each event. The last two columns report the patterns of interest (π_1 and π_2 in the table) and the number of occurrences of these patterns in the historical logging data. Specifically, the latter is the number of traces in the historical logging data for which there exists a prefix that leads to a state subsumed by the given state and in which the pattern occurred.

3.3   Evidence-Enhanced Prediction

This step takes as input *(i)* the prediction model built in the previous step, *(ii)* the set of patterns to predict, and *(iii)* the (partial) trace(s) corresponding to the running process execution(s), and returns (for each trace) a prediction on the patterns of interest. Each prediction should account for the amount of evidence for a given pattern in a given state. To this end, we employ principles of Subjective Logic [12]. This is an opinion algebra commonly used in the context of online communities, where users have to decide whether to interact with another user to achieve some goal. In assessing the trust level between users who do not know each other beforehand, an *opinion* is computed for each user based on his past interactions with other users in the community. Opinions are defined as follows:

Definition 6. *An opinion x about a proposition P is a tuple $x = (x_b, x_d, x_u)$ where x_b represents the belief that P is provable (belief), x_d the belief that P is disprovable (disbelief) and x_u that P is neither provable or unprovable (uncertainty). The components of x satisfies $x_b + x_d + x_u = 1$.*

Opinions are computed from evidence. Let p, n be the amount of evidence that supports the proposition and contradicts the proposition, respectively. The opinion x on the proposition P is computed as follows:

$$x_b = \frac{p}{p+n+c}; \quad x_d = \frac{n}{p+n+c}; \quad x_u = \frac{c}{p+n+c} \tag{2}$$

where $c > 0$ is a constant that represents the minimum amount of evidence required to form an opinion.[2]

For our purposes, we formulate the proposition P as an assertion on the occurrence of a given pattern π. In this setting, the evidence to compute opinions regarding P is derived by the prediction model. More precisely, given a state, p represents the number of traces leading to that state in which π occurred, whereas n is the number of traces leading to that state in which the pattern did not occur. Accordingly, x_b represents the belief that π will occur in a given state, x_d the disbelief that π will occur, and x_u the uncertainty about the computed opinions.

---

[2] Note that the original formulation of Subjective Logic in [12] assumes $c = 2$. However, later work [20] has shown that c can be a generic constant, which can be determined by the context.

Given an opinion x on the occurrence of pattern π for a trace σ, we compute a prediction from x at run-time. In doing so, two main aspects should be taken into account. First, we need to define a "proper" value for the minimum amount of evidence, that allows us to discard all those predictions that are "too much" uncertain. Secondly, we need to determine a suitable sensitivity for the prediction, since we want an alert to be raised only when we are "reasonably sure" that an undesired behavior is about to occur. In other words, we expect a positive answer only when belief x_b is "reasonably larger" than disbelief x_d.

It is worth noting that the concrete instance of what is a "reasonable" amount of minimum evidence and of how much x_b has to overcome x_d to get a positive answer are domain-dependent decisions. Therefore, we model these notions in form of parameters, which can be set by the decision maker based on her needs and preferences.

More precisely, given an opinion $x = (x_b, x_d, x_u)$ denoting the occurrence of a pattern π given a running execution σ w.r.t. a prediction model m, we compute the prediction as follows:

$$
pred(x) = \begin{cases}
Unpredicted & \text{if } x_u > x_b \wedge x_u > x_d \\
Yes & \text{if } (x_u < x_b \vee x_u < x_d) \wedge x_b > \alpha \cdot x_d \\
No & \text{otherwise}
\end{cases}
$$

where α is the sensitivity threshold (i.e., the required gap between belief and disbelief).

4   Experiments

This section describes the evaluation of our approach. In detail, we are interested in answering the following research questions:

RQ1: To what extent do the parameters c and α affect accuracy of the predictions?

RQ2: To what extent does the adopted data discretization strategy affect the accuracy of predictions?

RQ3: Is the accuracy of the predictions obtained with our approach in line with the one provided by previous predictive process monitoring techniques?

The first research question aims to investigate the impact of the two parameters used by our approach – α and c – on the quality of predictions. **RQ2** aims to provide insights on the differences between "local" and "global" data discretization in terms of prediction quality. In the first case, we perform data discretization w.r.t. each pattern individually. This way, we obtain a preprocessed log for each pattern, which is used to make predictions for the corresponding pattern only. In the second case, attributes are discretized by considering the occurrences of all patterns at once. Finally, **RQ3** aims to compare our approach with existing techniques. In order to address these questions and conduct an insightful experiment on realistic logs, we need datasets that allow us to evaluate the correctness of our predictions, i.e., how close the outcome of the proposed technique is to the desiderata, and gain meaningful insights about possible reasons underlying unexpected good/bad results. In other terms, we need event logs with data, labelled based on the occurrence of some patterns that involve also data, and some domain knowledge about these patterns. As far as we know, real-life publicly available datasets with these characteristics do not exist. Indeed, publicly available real-world event logs (e.g., https://data.4tu.nl/repository/collection:event_logs) typically involve complex data, for which little or no domain knowledge on the generating process is available, thus making it challenging to assess the correctness and relevance of the derived insights. We hence made a serious effort in order to generate a realistic dataset starting from one of these real-life logs (BPI2012), discovering the corresponding process model (see Fig. 1), injecting realistic data (see details below) on the simulated event log and labelling traces according to the occurrence of patterns also involving data.

We evaluated our approach against the results obtained by applying the clustering-based approach in [9], which explicitly takes into account the amount of evidence to compute predictions. In this approach, traces with a similar control-flow are first grouped together, and a classifier is trained on each cluster. The most suitable cluster (and, hence, classifier) is chosen at run-time to classify the current sample. If there is not enough evidence to make a decision, i.e., if the support of the (cluster of) traces representing the same state of the current one is below a user-defined threshold, no prediction is provided.

The following subsections describe the implementation and the parameters setting for each tested technique, the metrics used for the evaluation and the obtained results.

4.1   Experiments Settings

Dataset generation: To design a synthetic experiment exhibiting the complexity of real-word scenarios, we choose for our experiments a loan management process derived from the event log made available for the BPI2012 challenge based on several previous work (see Fig. 1). Based on this model, we generated a synthetic event log using CPNTools (http://cpntools.org/), a widely used tool for Petri nets editing and simulation, by setting for each pattern a probability of occurrence of 20%. We exploited the simulation options available in CPN to deal with changes in the control-flow, e.g., delaying the completion of *fraud checking*; while we developed a script in Java for the generation of values for the *amount* and *duration* attributes. To set possible values of the *amount* attribute, we collected the amount values from the BPI2012 log; then, for each trace in our event log, we randomly selected one of these values, setting a probability of 70% of selecting values higher than 10000, which is the threshold set for the fraud checking pattern. To capture the old applications resuming pattern, first we introduce a waiting time in between each pair of consecutive events, randomly choosing between an interval from 4 to 100 hours. Then, in those traces in which an application was resumed, we set a probability of 80% of increasing the timestamp of activity *a_registered* by 31 days, to ensure a reasonable number of cases in which the pattern occurred. The final support for each pattern derives by the combination of the support of the changes performed within

CPNTools and the changes performed by our script. More precisely, we obtained the following support values: 22.85% for the old applications resuming pattern; 14.81% for the delayed fraud check pattern; 29.86% for the multiple negotiations pattern; and, finally 21.62% for the applications resuming pattern. For the sake of simplicity, hereafter we refer to these patterns as *duration, fraud, negotiation* and *resumed* patterns respectively. It is worth noting that, by construction, we also generated traces involving partial patterns, e.g., patterns in which an old application was resumed but within an acceptable time window. This provides a realistic scenario for the log as we do expect a certain behavior to be undesirable only under certain conditions.

Data discretization: For data discretization, we used the supervised discretization approach provided by the Weka J48 decision tree implementation of the C4.5 algorithm [17]. Specifically, we looked at the discretized intervals returned by the decision tree algorithm by taking into account not only the continuous variables but also the categorical ones. In detail, we encoded the execution traces with the *last-payload* encoding [14] and we trained the decision tree. We then inspected the resulting decision tree and extracted the intervals of the continuous variables, whenever they have enough discriminative power with respect to the specific security pattern. We set the confidence threshold to 0.8 and the support threshold to 50.[3] We tested both the local and global discretization strategy.

Subjective logic classifier: We varied α between $[1, 2]$, with steps of 0.1. For each value of α, we tested three values of c, i.e. 2, 10 and 50. We did not consider higher values, since we observed a significant worsening of the classification performances already when setting the minimum amount of evidence to 20 traces.

Clustering-based approach: For the clustering-based approach, we used the K-means algorithm with 18 clusters[4] as clustering technique for grouping together execution traces with a similar control flow and the decision tree as classifier to be trained with data payload for the classification. As for classification thresholds, we varied the confidence threshold γ in the interval 0.5 and 0.9 and support threshold ρ in the set $\{2, 20, 50\}$.

Evaluation metrics: We evaluated our results along two dimensions:

- *Classification accuracy.* We evaluate this dimension both in terms of the standard classifier *Accuracy*, which is the percentage of samples correctly classified in the dataset, and *F1 measure* (F1 hereafter). The latter is a metric widely used when addressing imbalanced datasets, where one class is more represented than the other. F1 balances the *precision* of the classifier, intended as the exactness of the predictions, and its *recall*, intended as the completeness of the results. Both accuracy and F1 range between 0 (minimum) and 1 (maximum).

---

[3] This setting allows obtaining a reasonable number of discretization intervals.

[4] We applied a grid search and selected the number of clusters that optimizes the accuracy [8].

– *Failure rate.* A classifier usually does not provide a prediction when too little evidence is available. We measure failure rate (FR) as the percentage of unclassified samples in the dataset. Given the same values for classification accuracy, the best classifier is the one that achieves the lowest failure rate.

We evaluated the performance of the classifiers by means of a 10-fold cross validation.

4.2   Results

In the following we discuss the results related to each research question.

RQ1: Figure 4 shows the values of accuracy, F1 and FR while varying α and c. The approach performs well in terms of accuracy for all tested configurations. It achieved an accuracy around 90% for all patterns with the exception of the *negotiation* pattern, for which we obtained an accuracy around 75%. Varying the c parameter does not seem to impact the accuracy metric. For F1, we still obtained quite good results for most of the patterns, around 70–75%, although we observe degradation in the results while increasing α and c. However, the approach scored quite poorly in terms of F1 as regards the *negotiation* pattern, for which we obtained values between 0.4 and 0.5. As a general trend, we can observe an evident worsening of the performance in terms of F1 while α and, in particular, c increase. As regards the FR, the approach performs well for $c = 2$ (predictions are missing for at most the 0.04% of the samples), while performance worsens for higher values of c. In particular, for $c = 50$, relevant portions of the samples (around 30–40% on average) are missed. Note that, as expected, the FR is not affected by variations of α.

RQ2: By applying local discretization, we identified two classes for attribute *amount* (for the *fraud* pattern) and six classes for *duration* (for the *duration violation* pattern). Comparing the thresholds used by construction for generating the log against the closest values delimiting the corresponding discretization classes, we can observe that the results are in line with the construction parameters. For instance, for *amount* the construction threshold is 10000 and the delimiter of the discretization class, obtained from the decision tree, is 10030. Moreover, by comparing the thresholds identified with the discretization against the actual data in the log, we can observe that the identified thresholds actually fit the trace labelling of the log. For instance, when *amount* is lower than or equal to 10030, the *fraud* pattern never occurs (the smallest *amount* value for which a violation occurs is 10071). With the global discretization configuration, we obtain intervals only slightly different from the ones of the local discretization. This preliminary and qualitative analysis allows us to assess that the returned classes are reasonable with respect to the criteria used for the log construction and with respect to the actual log.

The prediction results obtained for the local and global strategies are similar with respect to all three metrics (Fig. 4). An exception is represented by the *duration* pattern, for which the locally discretized log leads to better values in

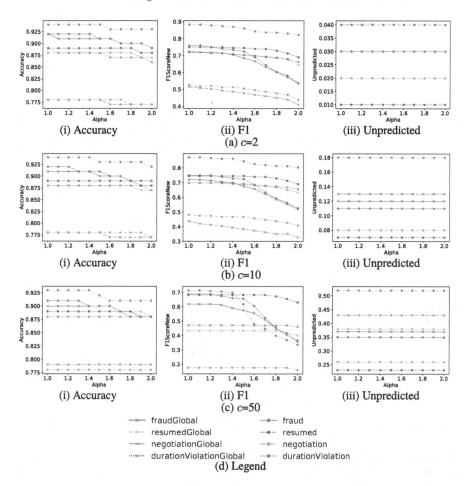

Fig. 4. Results of our approach for c equal to (a) 2, (b) 10 and (c) 50.

terms of accuracy and F1; however, this comes at the cost of a much higher rate of unpredicted samples.

RQ3: A comparison between our approach (SL) and the clustering-based approach is reported in Table 1 for the local discretization strategy and in Table 2 for the global discretization strategy. For each pattern, we report the configuration of parameters that optimizes each metric, along with the values of the other metrics for that configuration. For example, the first group of columns shows the configuration that optimizes accuracy.

For locally discretized data, the two approaches provided comparable performance in most of the cases. Overall, the clustering approach seems to return slightly better values for accuracy and F1 when considering the corresponding best configurations; however, this often comes with a much higher failure rate, in particular for the *fraud* and *negotiation* patterns. When considering the configu-

Table 1. Results on the locally discretized data

Pattern	Approach	Best accuracy				Best $F1$				Best FR			
		Par	Acc	$F1$	FR	Par	Acc	$F1$	FR	Par	Acc	$F1$	FR
fraud	SL	$\alpha = 1$ $c = 2$	0.92	0.76	0.03	$\alpha = 1$ $c = 2$	0.92	0.76	0.03	$\alpha = 1$ $c = 2$	0.92	0.76	0.03
	Clustering	$\gamma = 0.9$ $\rho = 50$	0.98	0.45	0.61	$\gamma = 0.9$ $\rho = 50$	0.98	0.45	0.61	$\gamma = 0.5$ $\rho = 50$	0.87	0.44	0.00
duration	SL	$\alpha = 1$ $c = 2$	0.94	0.88	0.04	$\alpha = 1$ $c = 2$	0.94	0.88	0.04	$\alpha = 1$ $c = 2$	0.94	0.88	0.04
	Clustering	$\gamma = 0.9$ $\rho = 20$	0.97	0.91	0.09	$\gamma = 0.9$ $\rho = 20$	0.97	0.91	0.09	$\gamma = 0.5$ $\rho = 2$	0.95	0.90	0.00
negotiation	SL	$\alpha = 1$ $c = 2$	0.78	0.53	0.02	$\alpha = 1$ $c = 2$	0.78	0.53	0.02	$\alpha = 1$ $c = 2$	0.78	0.53	0.02
	Clustering	$\gamma = 0.9$ $\rho = 20$	1.00	0.70	0.93	$\gamma = 0.9$ $\rho = 20$	1.00	0.70	0.93	$\gamma = 0.5$ $\rho = 20$	0.75	0.40	0.00
resumed	SL	$\alpha = 1$ $c = 2$	0.89	0.75	0.01	$\alpha = 1$ $c = 2$	0.89	0.75	0.01	$\alpha = 1$ $c = 2$	0.89	0.75	0.01
	Clustering	$\gamma = 0.9$ $\rho = 20$	0.98	0.90	0.21	$\gamma = 0.9$ $\rho = 20$	0.98	0.90	0.21	$\gamma = 0.5$ $\rho = 50$	0.91	0.80	0.00

rations that optimize failure rate, the clustering approach was able to classify all samples, even though our approach achieves a failure rate close to 0 with comparable results in terms of accuracy and F-measure. Results related to the globally discretized data show trends for accuracy and F1 similar to those observed in the locally discretized data. However, the clustering approach performs much worse than our approach in terms of failure rate for all configurations; indeed, failure rate is never below 23%, against the 0.04% achieved by our approach.

Discussion: Our approach scores fairly well for most of the patterns, although performance is worse for the *negotiation* pattern. This likely happens because this pattern contains a loop. Indeed, different number of loop iterations are modeled as different states in the predictive model, so that some states do not have enough support for being accounted in the prediction, leading to miss some positive samples. In some cases, e.g. for $\alpha = 1$, $c = 50$, no positive samples were found, thus resulting in F1 to be undefined. The results show that low values of α and c typically provide better results. This is not unexpected; higher the thresholds are, higher the probability of missing true positives is, which explains the performance worsening. The approach seems, however, to be much more sensitive with respect to c than to α. While the worsening in performance is negligible when varying α (with the exception of the *duration* pattern), we observed a worsening both in terms of *F1* and, especially, in terms of failure rate when increasing c. According to these results, it seems advisable setting low thresholds for the minimum amount of evidence, when possible. As regards RQ2, even though the final outcome is clearly affected by the employed discretization strategy, the results seem to suggest that adopting a local or a global strategy does not have a significant impact on the performance. For RQ3, we observed that the performance provided by our approach is similar to the one provided by the clustering-based approach. The latter performs sometimes better in terms of accuracy and F1 but misses a higher percentage of samples in most of the

Table 2. Results on the globally discretized data

Pattern	Approach	Best accuracy				Best $F1$				Best FR			
		Par	Acc	$F1$	FR	Par	Acc	$F1$	FR	Par	Acc	$F1$	FR
fraud	SL	$\alpha=1$ $c=2$	0.92	0.72	0.03	$\alpha=1$ $c=2$	0.92	0.72	0.03	$\alpha=1$ $c=2$	0.92	0.72	0.03
	Clustering	$\gamma=0.9$ $\rho=2$	1.00	0.98	0.96	$\gamma=0.9$ $\rho=2$	1.00	0.98	0.96	$\gamma=0.5$ $\rho=2$	0.89	0.35	0.23
duration	SL	$\alpha=1$ $c=2$	0.88	0.72	0.04	$\alpha=1$ $c=2$	0.88	0.72	0.04	$\alpha=1$ $c=2$	0.88	0.72	0.04
	Clustering	$\gamma=0.7$ $\rho=2$	1.00	0.93	0.76	$\gamma=0.9$ $\rho=2$	1.00	0.99	0.96	$\gamma=0.5$ $\rho=2$	0.91	0.57	0.23
negotiation	SL	$\alpha=1$ $c=50$	0.79	–	0.43	$\alpha=1$ $c=2$	0.78	0.51	0.04	$\alpha=1$ $c=2$	0.78	0.51	0.04
	Clustering	$\gamma=0.9$ $\rho=50$	0.98	0.22	0.96	$\gamma=0.9$ $\rho=2$	0.96	0.75	0.96	$\gamma=0.5$ $\rho=2$	0.83	0.59	0.23
resumed	SL	$\alpha=1$ $c=2$	0.89	0.72	0.03	$\alpha=1$ $c=2$	0.89	0.72	0.03	$\alpha=1$ $c=2$	0.89	0.72	0.03
	Clustering	$\gamma=0.8$ $\rho=20$	0.99	0.50	0.84	$\gamma=0.9$ $\rho=2$	0.98	0.75	0.96	$\gamma=0.5$ $\rho=2$	0.88	0.50	0.23

tested configurations. This is because the clustering-based approach uses stricter constraints and, in particular, relies on the classifier confidence to decide whether a prediction should be made. The similarity between the performance of the two approaches seems to suggest that relaxing the assumption on the confidence does not lead to significantly worse performance in terms of *accuracy* and *F1* while providing a lower failure rate. Moreover, the performance of the clustering-based approach seem to be more sensitive to the discretization strategy than our approach.

Threats to validity: One of the threats to the *external* validity of the evaluation is the application of the approach only to synthetic data. The use of more logs, including a real-life one, would clearly allow for more general results. However, such a threat is mitigated by the fact that the considered log was generated by simulating a realistic and widely known model, with a realistic number, type and range of data attributes. A second threat to the external validity is the choice of the investigated patterns. Also in this case the threat is mitigated by the fact that the chosen patterns are realistic for the considered scenario.

5   Related Work and Concluding Remarks

Predictive business process monitoring has received an increasing attention in the last years [15,16]. Existing approaches can be grouped in three main categories based on the aim of the prediction: *(i)* approaches that aim to predict the *remaining execution time* of running process instances, e.g. [19,22]; *(ii)* approaches that aim to predict the *next activity* to be executed, e.g. [5]; *(iii)* the so-called *outcome-oriented* approaches, which classify ongoing executions according to a given set of possible categorical outcomes [7,9,14,21]. Our work is related to the third group, since we predict the value of an indicator (i.e., the occurrence of a given pattern) for each running execution.

Within the outcome-oriented approaches, some works focus on predictions and recommendations to reduce risks [6,7,15]. For example, in [6,7], the authors present a technique to support process participants in making risk-informed decisions with the aim of reducing process failures, by considering process executions both in isolation [7] and propagating information about risks to similar running instances [6]. In [15], three different approaches for the prediction of process instance constraint violations are investigated: machine learning, constraint satisfaction and QoS aggregation. Our work, by making predictions on the occurrence of undesired behavior, is related to this group of works, although the focus is slightly different.

Moreover, our approach is close to those applying a *lossless* encoding strategy (e.g., [14]), which is an encoding that allows recovering the original trace. Since a lossless encoding leads to prefixes of different length, a common strategy adopted by lossless outcome-oriented approaches to employ classification techniques consists in splitting the set of prefixes in buckets, training a classifier for each of them. Different strategies have been explored to build the set of prefix buckets. For example, Lakshmanan et al. [13] build a classifier for each prefix length. Di Francescomarino et al. [9] exploit trace clustering techniques to group similar traces, building a classifier for each cluster. Leontjeva et al. [14] build a classifier for each state in a process model. Once the traces have been properly encoded and the prefixes have been grouped, well-known classification techniques are employed. Compared to previous work, our approach exploits a single predictive model, without requiring grouping prefixes and training multiple classifiers. Moreover, our approach provides analysts with a simple and intuitive mechanism to set both the reliability and sensitivity of predictions. To the best of our knowledge, sensitivity aspects have been mainly neglected by previous approaches; to support such a dimension, one should delve into the classification model and change the logic with which predictions are provided, which is not trivial and not always possible in existing approaches.

In this work, we introduced an approach based on Subjective Logic for predicting the occurrence of undesired behaviors in running process executions. The approach allows the process analyst to customize both the reliability and sensitivity desired for the prediction. This makes our predictive process monitoring system suitable for scenarios in which reacting to undesired behaviors might be costly or complex, as the system can raise an alert only in the presence of strong evidence supporting the prediction. The evaluation of the approach showed that the approach performed well both in terms of classification performance and effectiveness, obtaining results mostly comparable with the tested competitor and often leading to a significant reduction of the failure rate.

In future work, we plan to perform a more exhaustive set of experiments by considering real-world datasets as well as testing other discretization techniques. Moreover, we plan to investigate and develop solutions tailored to deal with the actor dimension. This dimension is necessary to predict behaviors involving actor-related constraints, such as separation/binding of duties, which cannot be handled by standard discretization techniques.

Acknowledgements. This work is funded by the ITEA3 project APPSTACLE (15017).

References

1. Sarbanes-Oxley act of 2002. public law 107–204 (116 statute 745), United States senate and house of representatives in congress (2002)
2. Adriansyah, A., Buijs, J.C.A.M.: Mining Process Performance from Event Logs: The BPI challenge 2012. BPM Center Report BPM-12-15 (2012). BPMcenter.org
3. Adriansyah, A., van Dongen, B.F., Zannone, N.: Controlling break-the-glass through alignment. In: Proceedings of SocialCom, pp. 606–611. IEEE (2013)
4. Alizadeh, M., de Leoni, M., Zannone, N.: Constructing probable explanations of nonconformity: a data-aware and history-based approach. In: Proceedings of Symposium Series on Computational Intelligence, pp. 1358–1365. IEEE (2015)
5. Breuker, D., Matzner, M., Delfmann, P., Becker, J.: Comprehensible predictive models for business processes. MIS Q. **40**(4), 1009–1034 (2016)
6. Conforti, R., de Leoni, M., La Rosa, M., van der Aalst, W.M.P., ter Hofstede, A.H.M.: A recommendation system for predicting risks across multiple business process instances. Decis. Support Syst. **69**(C), 1–19 (2015)
7. Conforti, R., de Leoni, M., La Rosa, M., van der Aalst, W.M.P.: Supporting risk-informed decisions during business process execution. In: Salinesi, C., Norrie, M.C., Pastor, Ó. (eds.) CAiSE 2013. LNCS, vol. 7908, pp. 116–132. Springer, Heidelberg (2013). https://doi.org/10.1007/978-3-642-38709-8_8
8. Di Francescomarino, C., Dumas, M., Federici, M., Ghidini, C., Maggi, F.M., Rizzi, W.: Predictive business process monitoring framework with hyperparameter optimization. In: Nurcan, S., Soffer, P., Bajec, M., Eder, J. (eds.) CAiSE 2016. LNCS, vol. 9694, pp. 361–376. Springer, Cham (2016). https://doi.org/10.1007/978-3-319-39696-5_22
9. Di Francescomarino, C., Dumas, M., Maggi, F.M., Teinemaa, I.: Clustering-based predictive process monitoring. IEEE Trans. Serv. Comput. 1 (2017)
10. El Kharbili, M.: Business process regulatory compliance management solution frameworks: a comparative evaluation. In: Proceedings of Asia-Pacific Conference on Conceptual Modelling, pp. 23–32. Australian Computer Society (2012)
11. Genga, L., Alizadeh, M., Potena, D., Diamantini, C., Zannone, N.: Discovering anomalous frequent patterns from partially ordered event logs. J. Intell. Inf. Syst. 1–44 (2018)
12. Jøsang, A.: Subjective Logic - A Formalism for Reasoning Under Uncertainty. Artificial Intelligence: Foundations, Theory, and Algorithms. Springer, Cham (2016)
13. Lakshmanan, G.T., Duan, S., Keyser, P.T., Curbera, F., Khalaf, R.: Predictive analytics for semi-structured case oriented business processes. In: zur Muehlen, M., Su, J. (eds.) BPM 2010. LNBIP, vol. 66, pp. 640–651. Springer, Heidelberg (2011). https://doi.org/10.1007/978-3-642-20511-8_59
14. Leontjeva, A., Conforti, R., Di Francescomarino, C., Dumas, M., Maggi, F.M.: Complex symbolic sequence encodings for predictive monitoring of business processes. In: Motahari-Nezhad, H.R., Recker, J., Weidlich, M. (eds.) BPM 2015. LNCS, vol. 9253, pp. 297–313. Springer, Cham (2015). https://doi.org/10.1007/978-3-319-23063-4_21
15. Metzger, A., et al.: Comparing and combining predictive business process monitoring techniques. IEEE Trans. Syst. Man Cybern. **45**(2), 276–290 (2015)

16. Mrquez-Chamorro, A.E., Resinas, M., Ruiz-Corts, A.: Predictive monitoring of business processes: a survey. IEEE Trans. Serv. Comput. 1–18 (2017)
17. Quinlan, R.: C4.5: Programs for Machine Learning. Morgan Kaufmann Publishers, Burlington (1993)
18. Ramezani, E., Fahland, D., van der Aalst, W.M.P.: Where did i misbehave? Diagnostic information in compliance checking. In: Barros, A., Gal, A., Kindler, E. (eds.) BPM 2012. LNCS, vol. 7481, pp. 262–278. Springer, Heidelberg (2012). https://doi.org/10.1007/978-3-642-32885-5_21
19. Rogge-Solti, A., Weske, M.: Prediction of business process durations using non-markovian stochastic petri nets. Inf. Syst. **54**, 1–14 (2015)
20. Skoric, B., de Hoogh, S.J.A., Zannone, N.: Flow-based reputation with uncertainty: evidence-based subjective logic. Int. J. Inf. Secur. **15**(4), 381–402 (2016)
21. Teinemaa, I., Dumas, M., La Rosa, M., Maggi, F.M.: Outcome-oriented predictive process monitoring: review and benchmark. arXiv:1707.06766 (2017)
22. van der Aalst, W.M., Schonenberg, M.H., Song, M.: Time prediction based on process mining. Inf. Syst. **36**(2), 450–475 (2011)

Counterfactual Reasoning for Process Optimization Using Structural Causal Models

Tanmayee Narendra[1], Prerna Agarwal[2(✉)], Monika Gupta[2],
and Sampath Dechu[1]

[1] IBM Research AI, Bangalore, India
tanmayee.narendra@iiitb.org, sampath.dechu@in.ibm.com
[2] IBM Research AI, New Delhi, India
{preragar,gupmonik}@in.ibm.com

Abstract. Business processes are complex and involve the execution of various steps using different resources that can be shared across various tasks. Processes require analysis and process owners need to constantly look for methods to improve process performance indicators. It is non-trivial to quantify the improvement of a proposed change, without implementing or conducting randomized controlled trials. In several cases, the cost and time for implementing and evaluating the benefits of these changes are high. To address this, we propose a principled framework using Structural Causal Models which formally codify existing cause-effect assumptions about the process, control confounding and answer "what if" questions with observational data. We formally define an end to end methodology which takes process execution logs and specified BPMN model as inputs for structural causal model discovery and for performing counterfactual reasoning. We show that exploiting the process specification for causal discovery automatically ensures the inclusion of subject matter expertise, and also provides an effective computational methodology. We illustrate the effectiveness of our approach by answering intervention and counterfactual questions on example process models.

Keywords: Structural causal model · Process optimization · What-if analysis · Counterfactual reasoning · Process redesign

1   Introduction

Business processes can be complex, involving multiple steps and resources [31]. While there is a constant need to improve process performance indicator (PPI) metrics, process re-engineering or re-allotment of resources does not necessarily guarantee improvements [24]. In addition, the decision of choosing which factor or aspect of a business process that needs to be changed is usually subjective and remains at the discretion of process owners or subject matter experts. The only concrete method of ascertaining that a proposed change to a process will definitely lead to improvements in PPI metrics, is to conduct randomized controlled

© Springer Nature Switzerland AG 2019
T. Hildebrandt et al. (Eds.): BPM 2019, LNBIP 360, pp. 91–106, 2019.
https://doi.org/10.1007/978-3-030-26643-1_6

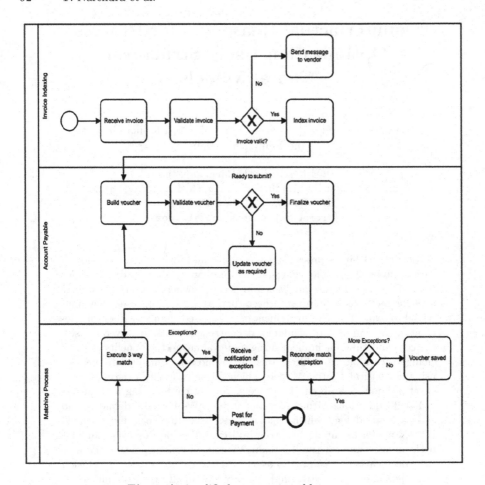

Fig. 1. A simplified account payable process

trials (RCTs) [20]. However, this is expensive and time-consuming, and hence inherently limits the number of changes that can be practically tested.

Business processes usually involve multiple actors. There are complicated interactions that occur between different components of the process. Causality provides a principled approach for modelling these complex correlations [4], and can be used to separate correlations from causation. Causal models control confounding (unmeasured variables) and help in avoiding Simpson's Paradox. They also help in mediation analysis, which involves answering what-if and retrospective questions that are expensive or impossible to answer otherwise.

1.1 Motivating Example

Consider the account payable process described in Fig. 1. Suppose a robot has been developed which does automatic voucher verification, and this robot is assumed to give 2x speed improvement for this activity as compared to a human agent. How can we give a quantifiable estimate of the improvement of some PPI, without deploying the robot in the field? This is a valuable question to answer in the field of Robotic Process Automation (RPA).

A suitable causal model helps in answering these 'what-if' kinds of questions using past, observational data about the process, such as event logs.

Table 1. Simple example to illustrate Simpson's paradox

	Overall	Digital	Scanned
EU Center	78% (173/350)	**93% (81/87)**	**73% (192/263)**
NA Center	**83% (289/350)**	87% (234/270)	69% (55/80)

Confounding in Business Processes. Consider the account payable process from earlier. In Table 1, the first column shows the overall percent of invoices processed within a day by European (EU) and North American (NA) centers. Notice that the time taken for ticket resolution within a day is 78% in the EU, while resolution within a day is 83% in NA. An analyst might stop here and conclude that ticket resolution is faster in NA processing center. However, let us group the data by the type of invoice, specifically, if it was scanned or digital. Now we see that EU Processing Centre is faster in both categories i.e., 93% for Digital and 73% for Scanned invoices (see row 1 of Table 1) as compared to NA Processing Centre where the percentage is 87% for Digital and 69% for Scanned invoices (see row 2 of Table 1). This reversal of behavior when the samples are grouped based on a confounding variable is known as Simpson's paradox. Here, the invoice type is a confounding variable. Therefore, when the invoices are grouped based on invoice type, reversal behavior is observed. The overall percentage of EU center is low because the EU receives more scanned invoices as compared to NA. Causal models provide a method to codify background knowledge about confounding variables, which in turn prevents process owners and analysts from reaching opposing conclusions from the same data.

1.2 Related Work

Causality and its theory has been successfully applied to a variety of other domains such as epidemiology [21], and problems in machine learning and computer science [1,4,32]. There has been some work on using causality to business processes [2,11,26,34].

Business process improvement is an active research area and has great practical relevance. Traditional methods of establishing process improvement are case studies and surveys [3,9,13,16,33], which are post hoc techniques. In [14], Process Improvement Patterns are introduced as best practices for improving the operational effectiveness of a process.

In this work, we are interested in answering what-if questions about the process, which is a pre-intervention analysis of the process. We provide a methodology that is relevant to scenario specific assessment of process improvement. We focus on causality as formalized in [17], where the focus is on answering questions on interventions and counterfactuals. To the best of our knowledge, this is the first attempt at using causality to answer what-if types of questions to business processes with the aim of better design.

The main contributions of our paper are as follows -

1. We introduce a principled framework with Structural Causal Models to answer intervention and counterfactual questions for a business process
2. We explain how the ordering implied by a BPMN model can be exploited for better discovery of the causal model for business processes
3. We illustrate the effectiveness of this method for answering what-if questions on illustrative processes.

2   Structural Causal Models for Business Processes

In this section, we introduce structural causal models in the context of business processes and describe how they can be used to answer 'what-if' questions in business processes.

In general, there are two parts to use causality to study a system -

1. **Causal discovery** or structure learning, which involves explicitly specifying cause-effect assumptions about the business process. These assumptions are encoded in the form of a Directed Acyclic Graph (DAG), where an edge (u, v) denotes that there is a direct causal relationship from $u \rightarrow v$.
2. **Causal inference**, which involves using data to answer what-if questions.

For the remainder of the paper, we will be considering the Structural Causal Model (SCM) framework [17,20] for our analysis. Structural Causal Models are designed to be able to answer three categories of questions [19]. This hierarchy is also referred to as the Ladder of Causation [18]. In increasing order of complexity, these categories are:

1. **Prediction** questions, such as "How many cases would be completed in an hour if I observed the inter-case arrival time to be x?"
2. **Intervention** questions, such as "How many cases would be completed in an hour if I made sure that *Task 1* took exactly 2 min to complete?"
3. **Counterfactual** questions, such as "Assuming everything else was the same, would my PPI metrics be better yesterday, had I ensured that *Task 2* was completed in y time?"

We formulate four major steps in the proposed framework - Data Aggregation and Random Variable Selection, Conditional Dependency Ordering, Causal Structure Learning and Answering Intervention Questions with Causal Inference. While the first three steps pertain to causal discovery, the last step is related to causal inference. Each of these components is explained in detail in the further sections. (See Fig. 2 for an overview).

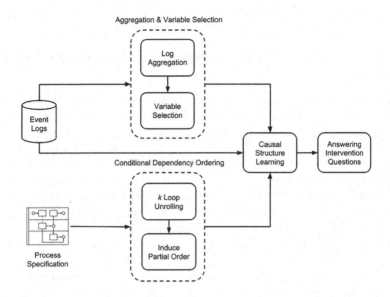

Fig. 2. Overview of building structural causal models for business processes

2.1 Data Aggregation and Random Variable Selection

In this step, relevant random variables about the business process are inferred from event logs. These random variables are constructed by aggregating the logs with respect to different aspects. Intuitively, since these variables will constitute the eventual Structural Causal Model that will be learned, it is important they are constructed keeping in mind the questions that need to be answered. Also, data aggregation must be done at the level of time granularity at which analysis is required.

Now, we will formally define how log aggregates are constructed. We borrow this definition of event logs from [30].

Definition 1 (Event Log). *An event log E consists of events where each event refers to a case, an activity, timestamp and event attributes (if applicable) such as actor and comments.*

From the perspective of causal modelling, event logs are considered as observational data. If specific experimental data (say, from Randomized Control Trials) about the business process was available, it would be considered intervention data. In this paper, we will consider causal modelling only from observational data.

Definition 2 (Log Aggregation). *Let the time frame of event logs be T_f and the duration of desired time granularity be T_s. Let T_f be divided into contiguous slices of duration T_s to obtain a set of time slices $T = \{t_1, ..., t_m\}$. Let the set of aspects of analysis be $\mathbb{X} = \{X_1, ..., X_k\}$. For a given event log E, log aggregate $\mathcal{L}$ is defined as*

$$\mathcal{L} = \left\{ \langle c_{X_i}^{t_j} \rangle \right\} \; \forall t_j \in T, \; \forall X_i \in \mathbb{X}$$

where $c_{X_i}^{t_j}$ denotes the number of process instances associated with aspect X_i at time slice t_j.

Informally, the set of aspects $\mathbb{X}$ must be chosen in such a way that all features of the business process that are relevant to questions that are desired being answered are captured. For example, if the goal was to answer 'what-if' questions at an hourly level of time granularity ($T_s = 60\,\text{min}$), the log aggregates must be constructed at 60-minute intervals. Also, if the objective was to answer intervention questions about organizational roles of human resources, then this aspect must be captured in the form of a random variable. In addition, $\mathbb{X}$ forms the set of random variables which will constitute the Structural Causal Model.

In the account payable process mentioned previously, one method of computing log aggregates is to count the number of process instances at each activity, at every hour of the day.

These log aggregates $\mathcal{L}$ form the basis for eventual empirical estimation of the causal structure and for answering of intervention and counterfactual questions.

2.2 Conditional Dependency Ordering

In this section, we will describe how any process specification with ordering semantics can be used to induce a partial order on the set of random variables $\mathbb{X}$. This step is important for two reasons - Firstly, by inducing an ordering from a process specification, we automatically include subject matter expertise into the model. This also ensures that the causal model learnt encapsulates the assumptions implied by the process. Secondly, this ordering has significant computational benefits for causal structure learning - a partial order of the random variables reduces the space of graphs to search.

Without loss of generality, we will assume that the process is specified in Business Process Modelling Notation (BPMN) for the rest of the paper. The graph structure of a BPMN diagram is utilized to specify a partial order.

We require a Directed Acyclic Graph (DAG) to induce a partial order. However, the associated BPMN model B may include cycles. We use k loop unrolling to eliminate cycles in the BPMN specification.

Definition 3. (k Loop Unrolling). *Given a directed cycle c in a directed graph G, loop unrolling involves repeating the vertices in c k times to generate a new graph $\hat{G}$ which is acyclic.*

k loop unrolling inherently results in loss of information, and k should be chosen in order to minimize this loss. See Fig. 3 for an illustration on k loop unrolling, where $k = 1$.

A DAG naturally induces a partial order on its vertices. Using this, we induce a partial order on the random variables themselves.

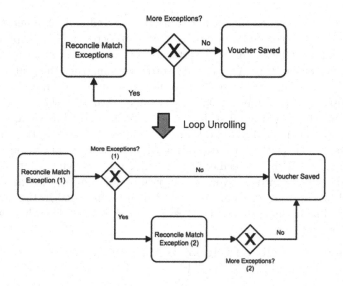

Fig. 3. Pictorial illustration of k loop unrolling

Definition 4 (Inducing Partial Order). *Each $X_i \in \mathbb{X}$ is associated with a node from the acyclic BPMN specification B^*, which in turn induces a partial ordering $\preceq$ on $\mathbb{X}$.*

Using this, we can induce a partial order of the aggregated random variables by using the BPMN specification. If we consider only the 'Invoice Indexing' swimlane of the account payable process, the partial order would be $X_{receive-invoice} \preceq X_{validate-invoice} \preceq X_{message-vendor}, X_{index-invoice}$.

2.3 Causal Structure Learning

In this section, we describe how log aggregates $\mathcal{L}$ and partial order $\preceq$ of a set of random variables $\mathbb{X}$ can be used to automatically learn the causal structure of the SCM. First, let us formally define Structural Causal Models, sometimes also called as Structural Equation Models.

Definition 5 (Structural Causal Models). *Consider a set of random variables $\mathbb{X} = \{X_1, ..., X_n\}$ pertaining to a given business process. The Structural Causal Model $\mathbb{S}$ for $\mathbb{X}$ is defined as the set of assignments*

$$X_i := f_i(\Pi_i, N_i) \; \forall i$$

where f_i is any function, $\Pi_i \subset \mathbb{X}$, $X_i \notin \Pi_i$ is the set of parents of X_i which have a direct cause-effect relationship to X_i, and N_i denotes noise. $N = \{N_i\}$ are required to be jointly independent.

These structural equations f_i are to be interpreted as assignments and are not bi-directional. Every SCM has an associated directed acyclic graph G, which is constructed as follows - make a vertex for every $X_i \in \mathbb{X}$ and draw an edge from every vertex in Π_i to X_i. In other words, we construct the graph is such a way that there are directed edges from a node's parents, to the node.

This DAG is also commonly called the causal structure [17,20]. The causal structure must be constructed such that it encodes expert knowledge and common sense assumptions about the business process. We encode this knowledge in two ways - First, in the form exploiting the partial order of the random variables $\mathbb{X}$ from the BPMN specification. Second, by specifying white-listed and black-listed edges. White-listed edges are required to be included in the causal graph, and black-listed edges are not allowed to be included in the graph. The goal of causal structure learning is to estimate the DAG G from log aggregates $\mathcal{L}$ and the partial order $\preceq$ of $\mathbb{X}$.

Definition 6 (Whitelisted Edges). *Whitelisted edges E_w are the set of edges that must be included in the SCM DAG G.*

Definition 7 (Blacklisted Edges). *For a given partial order $\preceq$ over a set of random variables $\mathbb{X}$, the set of blacklisted edges $E_b = \{(X_i, X_j)\ \forall X_j \preceq X_i\}$.*

In general, causal discovery from observational data is a hard problem, as the number of possible graphs for n random variables is $2^{\Omega(n^2)}$, which is super-exponential in n [7]. The problem becomes simpler if an ordering of variables is known, as the search space reduces to $2^{O(n \log n)}$ [27]. Alternatively, the problem of causal structure learning, given an ordering, can be reduced to variable selection in multivariate regression [5].

Causal structure learning algorithms can be divided into two broad categories based on their fundamental approach to the problem -

1. Score based, where a score is assigned to every DAG based on the goodness of fit with the data. Some score based algorithms are Hill-climb and Tabu search [15].
2. Constraint based, where the general approach is to use conditional independence tests to recover the causal DAG structure. Some constraint based algorithms are PC [6], GS [15], MMPC [28], IAMB [29], among many others.

While score-based algorithms are guaranteed to return DAGs, constraint based algorithms may return partially directed acyclic graphs (PDAGs), which are graphs which may contain undirected edges. The ordering based search algorithm described in [27] is a good candidate, as this is a score based algorithm that provides a heuristic for searching over the space of orderings, instead of the space of DAGs.

2.4 Answering Intervention and Counterfactual Questions

In this section, we describe how the learnt Structural Causal Model $\mathbb{S}$ can be used to estimate answers to intervention and counterfactual questions about the business process. First, let us formally define an intervention -

Definition 8 (Intervention). *For a SCM $\mathbb{S}$ defined over the set of random variables $\mathbb{X}$, an intervention is defined as setting a subset of variables $X \subset \mathbb{X}$ to a particular value x. This is denoted by $do(X = x)$.*

In this work, we will only consider atomic interventions which involve setting individual random variables to a particular value. Usually, we are interested in the estimation of some target quantity after an intervention. For example, if we were interested in answering the question - "How many cases would be completed in an hour if I made sure that *Task 1* took exactly 2 min to complete?", the intervention involves setting the time duration of *Task 1* to 2 min, and we are interested in seeing the effect of this change on the number of cases that reach completion in an hour. This target quantity is also called as outcome.

The first step towards answering intervention questions is graph surgery. Informally, this involves removing incoming edges to all vertices in the causal DAG that involves a variable which is being intervened on.

Definition 9 (Graph Surgery). *For an SCM $\mathbb{S}$ associated with DAG G, with intervention $do(X = x)$, the new graph G^* is generated by deleting all incoming edges for all $X_i \in X$. This is also called Graph Surgery.*

Next, we will show that for any SCM $\mathbb{S}$, the effect of any intervention on any other random variable can always be estimated.

Lemma 1 (Identifiability). *Consider any SCM $\mathbb{S}$. Let the intervention be denoted by $do(X = x)$ and let the target quantity of interest be denoted by Y. Then, $Pr(Y|do(X = x))$ can be determined from observational data. In other words, the causal effect is identifiable.*

Proof. Let the set of random variables which constitute $\mathbb{S}$ be denoted by $\mathbb{X}$. Let the new graph obtained after surgery be denoted by G^*.

For each node $X_i \in \mathbb{X}$, estimate X_i using $X_i := f_i(\Pi_i, N_i)$ where $X_i = x \; \forall X_i \in X$.

This reduces the problem of estimation of $Pr(Y|do(X = x))$ to a statistical estimation problem. Since all $X_i \in \mathbb{X}$ is known, $Pr(Y|do(X = x))$ can be determined.

Estimation can be treated as a purely statistical problem, and several approaches can be used. The choice of the approach depends on the nature of data, and statistical properties desired. Some common approaches that are commonly used in estimation are inverse probability weighting [10], propensity score matching [23], regression of the outcome on intervention variables, and double robust efficient methods [22, 25], including targeted maximum likelihood [12].

3 Experiments and Results

In this section, we illustrate the usefulness of our framework by applying it to a set of generic processes. The experimental setup and results are discussed in detail.

3.1 Experiment Setup

The following are the primary steps involved -

1. **Select BPMN:** We select a set of BPMN models which generalize well to model different use-cases. We start with a simple BPMN model with a sequence of tasks ending with just one end task node. We then increase the complexity of the BPMN model to include decision nodes and more tasks to illustrate the performance of our system. We further hypothesize that our system will be able to produce results on any BMPN model which consists of multiple decision nodes, sub-processes and end task nodes. We do not consider loops in BPMN models as of now. We consider BPMN specifications from BPMN 2.0 Test Cases (Models, Diagrams, Serializations) created by the BPMN Model Interchange Working Group (BPMN MIWG)[1]. This is a set of domain agnostic BPMN specifications which are used to test different BPMN tools. The exact models that were considered are shown in Figs. 4 and 5.[2]

2. **Generate logs:** We use the online BIMP tool [8] to simulate the event logs for the chosen BPMN specification. This tool gives us the flexibility to upload a BPMN model and specify attributes such as the number of resources, time duration etc. Also, it also allows the specification of the inter-arrival time of processes and the time duration of each task as a probability distribution. The parameters used for simulation in BIMP are specified in Table 3.

3. **Aggregate logs:** Aggregated logs are generated from simulated logs. Depending on the granularity of the analysis, we counted the number of process instances for each random variable for that duration. For experimental purposes, we used the duration as 1 hour to aggregate the logs. This gives us an estimate of the load on an hourly basis, which can be used to identify bottlenecks and delays in a process model.

4. **Estimate Causal Model:** To obtain the initial seed for the structure discovery algorithm to discover the final causal model, we first compute topological sort of the nodes to define the partial order. We then add the edges in the directions directed by the topological sort to obtain a partially complete graph. We then remove all those edges which violate the constraints. For example, there should not be an edge from task to resource. This is because a task may be dependent on a particular resource but the vice-versa is not true. After we obtain the initial seed, we use a score based modified Hill Climb Algorithm to discover the Causal Model. The Hill Climb algorithm traditionally performs 3 operations i.e., add, reverse and delete the edge. Since, we already have partial order defined by the topological sort, adding and reversing the edge while discovering the structure is no longer required. Hence, the

---

[1] GitHub - https://github.com/bpmn-miwg/bpmn-miwg-test-suite.

[2] Although there are several open datasets for business processes, to verify the correctness of a causal model, intervention data is required. To the best of our knowledge, there is no existing dataset which documents the effect of multiple interventions in the business process domain. Hence, we use simulated data to showcase the feasibility of our approach.

only operation required is deleting an edge. The causal model with the high-est score is considered as the final discovered model. The final model is then compared with the expected model to show its validity.

5. **Inference on the Causal Model:** The range of questions that can be answered by the SCM is limited by the expressiveness of the random variables that constitute it. For example, if log aggregation was done at a daily level, then intervention questions at the hourly level cannot be answered. Also, if resource utilization has not been captured as a random variable, then we cannot answer questions about this aspect. A non-exhaustive set of possible intervention and counterfactual questions are shown in Table 2.

   To perform inference, we assume parametric models at each node, and for simplicity, assume linear Gaussian models of the form

$$X_i = \sum_{X_j \in \Pi_i} \beta_j X_j + \mathcal{N}_i$$

where β_j are real constants, and $\mathcal{N}_i$ is a Gaussian distribution representing noise. This is the simplest possible setting.

Table 2. Possible set of intervention questions that can be answered by the B-SCM

Aspect	Question
Resource	How many tickets can I process in a day, if I reduce 5 resources in Activity A?
	Will the PPI improve if I increase per person utilization in Activity B?
Time	If the time for step A is reduced by half, how many tickets are terminated in RESOLVED state?
	If I reduce tickets processed to 10 per day, can I achieve desired SLA levels?

Fig. 4. BPMN A.1.0 from BPMN 2.0 Test cases

Table 3. Parameters for simulation on BIMP

A.1.0	
Inter arrival time	Uniform distribution, [0, 60]
Process instances	10000
Task 1	Uniform distribution, [5, 35]
Task 2	Uniform distribution, [10, 20]
Task 3	Uniform distribution, [5, 35]
A.2.0	
Inter arrival time	Uniform distribution, [0, 60]
Process instances	10000
Task 1	Uniform distribution, [5, 35]
Task 2	Uniform distribution, [10, 20]
Task 3	Uniform distribution, [5, 35]
Task 4	Uniform distribution, [10, 20]

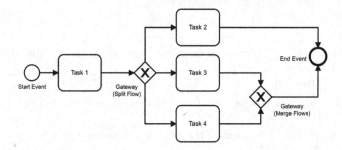

Fig. 5. BPMN A.2.0 from BPMN 2.0 Test cases

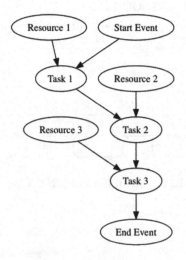

Fig. 6. SCM graph obtained for BPMN A.1.0 after structure learning

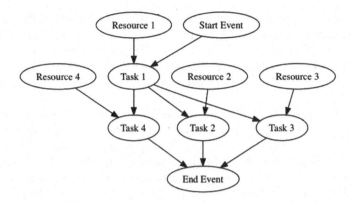

Fig. 7. SCM graph obtained for BPMN A.2.0 after structure learning

3.2 Results

Causal Structure Learning for Business Processes. Figures 6 and 7 show the SCM DAG discovered by structure learning algorithms. Notice that the diagram is very similar to a process control flow graph that is mined from traditional process mining algorithms. However, there are significant differences in the semantics of the nodes of the graph - for process flow diagrams, each node represents a state in the process, whereas, for the SCM DAG, each node represents an aggregated random variable. The SCM is intended to capture common sense knowledge and domain expertise of the business process, so it is important that the edges of the SCM DAG are in-line with what is already known about the process.

Causal Inference for Business Processes. We use linear Gaussian models at each node of the SCM. This is in turn used to estimate answers to intervention and counterfactual questions. From an implementation perspective, for each node X_i we fit a linear regression, with X_i as the target and Π_i as the predictors.[3] Refer Table 4 for details on the number of regression models and average accuracy for the BPMN specifications. Evidently, the average accuracy of the regression model drops with the increasing complexity of the business process.

Table 4. Performance metrics for regression models built at every node of SCM graph for BPMN Specifications

BPMN	Regressors	Average accuracy
A.1.0	4	99.1%
A.2.0	5	92.3%

---

[3] We tried three variants of linear regression - without regularization, and with L1 and L2 regularization. Empirically, the statistical model without regularization gave the best performance.

We will illustrate how an intervention question is answered with the SCM through an example. Let us consider the intervention where we make sure that for *Task 2*, 6 cases are processed every hour. We want to see the effect of this intervention on the hourly number of cases which reach the state *End Event*. If we denote *End Event* as X_{end} and *Task 2* as X_2, formally we want to estimate $\mathbb{E}\left[X_{end}|do(X_2 = 6)\right]$.

To answer this, we use the regression models at each node along with the SCM graph (Refer to Fig. 7). Let us assume that the logs were aggregated on an hourly basis, and the random variables represent the number of cases in that particular state or assigned to that resource at that hour. The intervention translates to setting *Task 2* to the constant 6. Also, all incoming edges to *Task 2* are removed. We are interested in the estimation of *End Event*, which has three predictors - *Task 4*, *Task 2* and *Task 3*. Since the value of *Task 2* is already fixed, we need to find values of *Task 3* and *Task 4*. To do this, we use the new graph (after removal of edges), and individual estimators of all other nodes.

Since we use linear Gaussian models, the estimation of the expected value of a particular node translates to plugging in the expected value of each of its parents into the structural equation.

How do we check the validity of the estimated quantity after intervention? We use the BIMP simulator to simulate the event logs for the process after the intervention. For the above example, this involves fixing the time duration taken for *Task 2* to 10 min. We then check the deviation between the predicted and simulated value. Table 5 shows the results for a selection of interventions.

Table 5. Selected set of interventions and deviations from simulated values

BPMN	Intervention	Value	Error
A.1.0	Resource 1	4	0.00095
A.1.0	Start Event	10	0.0067
A.2.0	Resource 1	3	0.0051
A.2.0	Task 2	6	0.016

4   Conclusion

In this work, we provide a principled framework to allow the answering of *what-if* questions about business process. In particular, we specify an end-to-end methodology where Structural Causal Models can be used to codify existing cause-effect assumptions, control confounding, and answer intervention and counterfactual questions. We illustrate the effectiveness of this framework on domain agnostic BPMN specifications.

There are several avenues for future work - Firstly, it might be worthwhile to explore if the deviation between the estimation and simulated quantity can be reduced by using more expressive Structural Causal Models. Here, we considered only linear Gaussian models; however, it might be possible to achieve better performance and generalization with more robust equations using Generalized Additive Models (GAM), Gaussian mixtures or neural networks. Second, the SCM can be made more expressive by adding more facets of a business process such as stakeholder engagement, effective communication between actors, organizational information, such as actor roles, and many more. Lastly, it is important to consider how this framework can be made usable and intuitive to process owners, and how these results of interventions can be effectively communicated to them.

References

1. Athey, S., Imbens, G.W.: Machine learning methods for estimating heterogeneous causal effects. Stat **1050**(5) (2015)
2. Baker, J., Song, J., Jones, D.R.: Closing the loop: an empirical investigation of causality in it business value (2017)
3. Borboudakis, G., Tsamardinos, I.: Towards robust and versatile causal discovery for business applications. In: Proceedings of the 22nd ACM SIGKDD International Conference on Knowledge Discovery and Data Mining, pp. 1435–1444. ACM (2016)
4. Bottou, L., et al.: Counterfactual reasoning and learning systems: the example of computational advertising. J. Mach. Learn. Res. **14**(1), 3207–3260 (2013)
5. Bühlmann, P., Peters, J., Ernest, J., et al.: Cam: causal additive models, high-dimensional order search and penalized regression. Ann. Stat. **42**(6), 2526–2556 (2014)
6. Colombo, D., Maathuis, M.H.: Order-independent constraint-based causal structure learning. J. Mach. Learn. Res. **15**(1), 3741–3782 (2014)
7. Finch, S.R.: Mathematical Constants, vol. 93. Cambridge University Press, Cambridge (2003)
8. Freitas, A.P., Pereira, J.L.M.: Process simulation support in BPM tools: the case of BPMN (2015)
9. Greco, G., Guzzo, A., Lupia, F., Pontieri, L.: Process discovery under precedence constraints. ACM Trans. Knowl. Discov. Data (TKDD) **9**(4), 32 (2015)
10. Hernán, M.A., Lanoy, E., Costagliola, D., Robins, J.M.: Comparison of dynamic treatment regimes via inverse probability weighting. Basic Clin. Pharmacol. Toxicol. **98**(3), 237–242 (2006)
11. Hompes, B.F.A., Maaradji, A., La Rosa, M., Dumas, M., Buijs, J.C.A.M., van der Aalst, W.M.P.: Discovering causal factors explaining business process performance variation. In: Dubois, E., Pohl, K. (eds.) CAiSE 2017. LNCS, vol. 10253, pp. 177–192. Springer, Cham (2017). https://doi.org/10.1007/978-3-319-59536-8_12
12. Van der Laan, M.J., Rose, S.: Targeted Learning: Causal Inference for Observational and Experimental Data. Springer, New York (2011). https://doi.org/10.1007/978-1-4419-9782-1
13. Limam Mansar, S., Reijers, H.A.: Best practices in business process redesign: use and impact. Bus. Process Manag. J. **13**(2), 193–213 (2007)
14. Lohrmann, M., Reichert, M.: Effective application of process improvement patterns to business processes. Softw. Syst. Model. **15**(2), 353–375 (2016)
15. Margaritis, D.: Learning Bayesian network model structure from data. Technical report. Carnegie-Mellon Univ Pittsburgh Pa School of Computer Science (2003)

16. zur Muehlen, M., Ho, D.T.: Service process innovation: a case study of BPMN in practice. In: Proceedings of the 41st Annual Hawaii International Conference on System Sciences (HICSS 2008), pp. 372–372. IEEE (2008)
17. Pearl, J.: Causality. Cambridge University Press, Cambridge (2009)
18. Pearl, J.: The eight pillars of causal wisdom. Lecture Notes for the UCLA WCE (2017)
19. Pearl, J.: Theoretical impediments to machine learning with seven sparks from the causal revolution. arXiv preprint arXiv:1801.04016 (2018)
20. Peters, J., Janzing, D., Schölkopf, B.: Elements of Causal Inference: Foundations and Learning Algorithms. MIT Press, Cambridge (2017)
21. Petersen, M.L., van der Laan, M.J.: Causal models and learning from data: integrating causal modeling and statistical estimation. Epidemiology (Cambridge, Mass.) **25**(3), 418 (2014)
22. Robins, J.M., Rotnitzky, A., Zhao, L.P.: Estimation of regression coefficients when some regressors are not always observed. J. Am. Stat. Assoc. **89**(427), 846–866 (1994)
23. Rosenbaum, P.R., Rubin, D.B.: The central role of the propensity score in observational studies for causal effects. Biometrika **70**(1), 41–55 (1983)
24. Satyal, S., Weber, I., Paik, H., Di Ciccio, C., Mendling, J.: AB-BPM: performance-driven instance routing for business process improvement. In: Carmona, J., Engels, G., Kumar, A. (eds.) BPM 2017. LNCS, vol. 10445, pp. 113–129. Springer, Cham (2017). https://doi.org/10.1007/978-3-319-65000-5_7
25. Scharfstein, D.O., Rotnitzky, A., Robins, J.M.: Adjusting for nonignorable dropout using semiparametric nonresponse models. J. Am. Stat. Assoc. **94**(448), 1096–1120 (1999)
26. Shook, C.L., Ketchen Jr., D.J., Hult, G.T.M., Kacmar, K.M.: An assessment of the use of structural equation modeling in strategic management research. Strateg. Manag. J. **25**(4), 397–404 (2004)
27. Teyssier, M., Koller, D.: Ordering-based search: a simple and effective algorithm for learning Bayesian networks. arXiv preprint arXiv:1207.1429 (2012)
28. Tsamardinos, I., Aliferis, C.F., Statnikov, A.: Time and sample efficient discovery of Markov blankets and direct causal relations. In: Proceedings of the ninth ACM SIGKDD International Conference on Knowledge Discovery and Data Mining, pp. 673–678. ACM (2003)
29. Tsamardinos, I., Aliferis, C.F., Statnikov, A.R., Statnikov, E.: Algorithms for large scale Markov blanket discovery. In: FLAIRS Conference, vol. 2, pp. 376–380 (2003)
30. Van Der Aalst, W.: Process Mining: Discovery, Conformance and Enhancement of Business Processes, vol. 2. Springer, Heidelberg (2011)
31. van der Aalst, W.M.P., ter Hofstede, A.H.M., Weske, M.: Business process management: a survey. In: van der Aalst, W.M.P., Weske, M. (eds.) BPM 2003. LNCS, vol. 2678, pp. 1–12. Springer, Heidelberg (2003). https://doi.org/10.1007/3-540-44895-0_1
32. Wang, Y., Liang, D., Charlin, L., Blei, D.M.: The deconfounded recommender: a causal inference approach to recommendation. arXiv preprint arXiv:1808.06581 (2018)
33. Wu, Y., Zhang, L., Wu, X.: On discrimination discovery and removal in ranked data using causal graph. In: Proceedings of the 24th ACM SIGKDD International Conference on Knowledge Discovery & Data Mining, pp. 2536–2544. ACM (2018)
34. Yusof, Z., Yusoff, W.F.W., Maarof, F.: Causality analysis in business performance measurement system using system dynamics methodology. In: AIP Conference Proceedings, vol. 1605, pp. 1201–1206. AIP (2014)

A Java-Based Framework for Case Management Applications

André Zensen[(✉)] and Jochen M. Küster[(✉)]

Bielefeld University of Applied Sciences, Bielefeld, Germany
{andre.zensen,jochen.kuester}@fh-bielefeld.de

Abstract. Case Management aims to support knowledge-intensive, flexible and non-routine processes. While modeling of Case Management applications is supported by the notation Case Management and Model Notation, the design and implementation of such applications can be realized using one of the currently available heavy-weight tool suites. In situations where such heavy-weight tool suites cannot be applied, a case management application must be implemented step by step from scratch. In this paper, we propose a framework to systematically create light-weight Case Management applications with CMMN execution semantics. Reusable elements are assembled into case blueprints and executed together with individual implementations. Case Management applications can be realized with less effort compared to a step by step approach. Developers can focus on business logic and supporting graphical user interfaces. As a proof of concept, we use our framework to implement part of a more complex CMMN case study.

1 Introduction

Knowledge work and knowledge-intensive processes gain importance especially in industrialized and information societies. While routine work and known procedures can be supported and automated by IT systems such as work flow systems, this is not necessarily the case for flexible and unstructured work. Yet this kind of work is more and more needed to adapt to new situations. Case Management (CM) aims to support knowledge-intensive, flexible and non-routine processes. The Case Management and Model Notation (CMMN) aims to become a standard notation for CM.

After requirements for a Case Management solution have been specified and a CMMN model has been created, one approach to create a CM application is to use a commercially available proprietary solution, such as by ISIS Papyrus [22] or IBM [15]. However, these often do not use a notation standard and come with the disadvantages of high costs and vendor lock-in. Alternatively, a complex BPMN engine with partial CMMN functionality could be used, such as Camunda [2], with the disadvantage of only partial CMMN support and part (the BPMN part) of the engine being not used at all. Another different approach is to create a CM application from scratch, e.g. by programming a solution directly in Java.

© Springer Nature Switzerland AG 2019
T. Hildebrandt et al. (Eds.): BPM 2019, LNBIP 360, pp. 107–124, 2019.
https://doi.org/10.1007/978-3-030-26643-1_7

However, due to the complexity of the CMMN semantics, such an approach is also costly and time consuming.

In this paper, we present a framework for creating light-weight applications based on CMMN. The idea is to enable the systematic realisation of case management applications based on CMMN models. The framework provides reusable building blocks based on a simplified CMMN structure. The framework itself makes use of proven architectural and design patterns, is flexible and adaptable and not restricted to a specific domain. A case management application can then be assembled by reusing building blocks of the framework for reoccurring functionality of any CM application and complementing them with additional components, such as specific UI components. Using the current framework prototype, an example application based on a case study has been implemented as a proof of concept.

The paper is structured as follows: Sect. 2 introduces concepts of CM and addresses problems of traditional and activity-centric process management. It also introduces CMMN elements and its execution semantics. Requirements for the framework are derived in Sect. 3. In Sect. 4 the overall architecture of the framework is introduced. Section 5 describes how our framework can be used to implement CM applications. Section 6 reports on an implementation of a case study. Related work is discussed in Sect. 7, before a conclusion and outlook on future research is given in Sect. 8.

2   Concepts and Modeling of Case Management Processes

Activity-centric process management with tasks along pre-defined paths and structures has its limits with regards to non-predetermined paths and high degrees of flexibility. Modeling such flexibility with traditional tools quickly leads to cluttered models [25, 26].

CM processes are handled as often long lasting cases, which are coordinated and handled by case workers in a collaborative fashion. The goal of a case is usually known, while the path leading there varies or cannot be pre-determined at all [23].

CM organizes activities around a case file, such as those of a legal or medical setting [3]. The roots of case management can be found in patient care: Depending on a patient's needs and state of health, different treatments and services are chosen for the patient to reach the overarching goal of improving the patient's health. How exactly this goal is achieved might not be known beforehand, e.g. a method of treatment might not be available or applicable to the case and unplanned treatments might become necessary at a later point [14].

Tasks are performed based on data and generate it. Which tasks are to be performed or necessary data to be collected is decided by the case workers. Task states change, e.g. information becomes available or updated and this leads to different decisions, tasks and (intermediate) goals to be achieved. Paths are thus not pre-determined from the beginning, but evolve as the case progresses over time. This approach is not restricted to healthcare domains: highly flexible processes or parts thereof can also be handled as a case across very different domains such as production [23, 26].

Case management aims to address four criteria found to be problematic in traditional work flow management systems and process management [25]:

Context tunnelling is avoided by making information available to all case workers and not restricting it to single tasks currently worked on.

Flexible paths evolve over time dependent on available information and states instead of being restricted by pre-determined paths and previously absolved tasks.

Division of labour is not restricted to certain authorised execute roles, but shared among participating case workers who have different roles.

Data and information can be added and edited regardless of specific activities and is thus not bound to a temporal order of tasks.

At least three approaches to case management have emerged, mainly differing in their degree of flexibility [7]: Adaptive Case Management (ACM) [9, 10, 19], Dynamic Case Management (DCM) [21] and Production Case Management (PCM) [5]. While ACM is the most flexible and aims to enable case workers to build their case on the fly as a "do it yourself" during execution, DCM enables case workers to adapt and expand on existing case structures which were set up before execution. The least flexible of the three approaches is PCM. It can be seen as a best of breed of previously enacted ACM or DCM cases, using templates of tested and proven flexible structures and partially known paths. These are then combined into a case structure before execution.

The Case Model and Management Notation (CMMN) 1.1 [4] is a declarative approach to process and case modeling. Its specification regards case management as a *"proceeding that involves actions taken regarding a subject in a particular situation to achieve a desired outcome"*. It can be regarded as a modeling notation closest to the characteristics of the DCM and PCM approaches, but also captures ACM characteristics [6]. Its structure and execution semantics are heavily based on the Guard Stage Milestone approach (GSM) [8] and the case handling paradigm described in [25].

Figure 1 shows two (simplified) CMMN models based on a case study conducted in [26]. Like its 'sibling-notation', the Business Process and Management Notation (BPMN) [1], CMMN offers a formalised graphical notation with defined execution semantics for its elements as well as an underlying specific markup language. The shown models express the high flexibility provided by CMMN with its decorators and sentry concept controlling the work flow. The top case is started by case workers, who can also trigger the *EventListener* **Cancel** to terminate the parent case. The required *CaseTask* **Create Specifications** in *Stage* **Create Technical Specifications** starts the bottom case **Create Technical Specifications**. Both cases automatically complete once all required elements are completed (signified by the exclamation mark at the bottom of *HumanTask* **Assemble Specifications**).

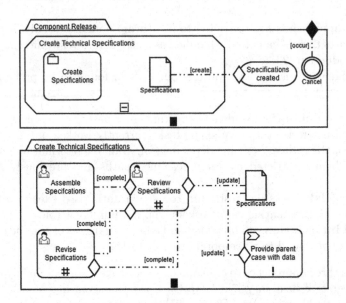

Fig. 1. Simple component release example CMMN model

The repetition decorators on *Human Tasks* **Review Specifications** and **Revise Specifications**) in the bottom case are used to model a loop structure. They evaluate a boolean *Property* of *CaseFileItem* **Specifications** which expresses whether or not the specifications have been approved or need to be revised and reviewed again.

Once the specifications are assembled, either in the context of work on the *Human Task* or by other means, one of the *EntrySentries* on *Human Task* **Review Specifications** is satisfied and transitions into an active state.

Apart from other *Sentries* observing state transitions of other elements and data in order to activate an element they are attached to, an *IfPart* controls entry to *ProcessTask* **Provide parent case with data**. Once the specifications are updated and have been approved, the *ProcessTask* is used to transfer data to the *CaseFileItem* of the parent case.

Then the *CaseFileItem* of the parent case undergoes its *create* transition, the *EntrySentry* on *Milestone* **Specifications created** is satisfied and the *Milestone* occurs, leading to a completion of the parent case.

The next chapter discusses requirements to build CM applications to support such cases.

3   Requirements for a Light-Weight Case Management Framework

CM applications can be characterized as distributed applications for collaborative work on long running cases. The application enables tasks depending on the overall context, data and state of a case instance. Case workers view and manipulate data to drive the case.

Requirements can be derived from CM characteristics [3,10] and our project work on component release processes as cases [26]. The requirements can be divided into base (1–6), CMMN specific (7–9) and those addressing problems of traditional process management (10–13). The list is not all-encompassing, but focus on minimum requirements for our framework to support light-weight, CMMN based CM applications.

Base requirements of the framework cover characteristics of CM:

REQ-1 *Offer a structured approach for implementing CM applications.* The framework needs to enable developers to implement CM applications in a systematic fashion.

REQ-2 *Remain light-weight.* Vendor-specific technology should be avoided, e.g. specific application server and database management system. Instead open standards should be used.

REQ-3 *Support case workers with graphical user interfaces.* Case workers need to view and edit data, often based on unstructured documents.

REQ-4 *Support multiple case workers with different roles.* A role system is required to restrict access to sensitive functions.

REQ-5 *Provide access for external systems.* Web-services are required for case workers accessing an application via web browsers and for external systems performing automated tasks, e.g. for call backs upon finishing the task.

REQ-6 *Support long running case instances.* A persistence mechanism is required to store and retrieve case instances which can last up to several years.

CMMN-specific requirements encompass building blocks and their behaviour:

REQ-7 *Provide foundational building blocks for a CM application.* The framework should provide classes representing CMMN elements, such as *Stage*, *HumanTask* and *ProcessTask*, in order to build case structures.

REQ-8 *Capture CMMN execution semantics and behaviour.* A common understanding of how the case instances behave needs to be established by basing behaviour of building blocks on the CMMN specification of execution semantics including those of decorators.

REQ-9 *Support the sentry concept.* Central to dynamic flows in CMMN are *Sentries*. The framework needs to support these to link elements together in order to influence states of elements, the availability and necessity of tasks to help guide case workers and to open possible pathways depending on element states and data values.

The following requirements address four problem areas of traditional process management, such as context tunnelling, rigid roles and path flexibility (see Sect. 2):

REQ-10 *Avoid context tunnelling.* All case workers should be able to access all information of a case instance to make better decisions.

REQ-11 *Enable flexible paths.* CMMN structures and execution semantics enable highly flexible and context dependent paths. *Milestones* further aid to focus on achieving a goal and what *can* be done instead of rigid paths focusing on what *should* be done.

REQ-12 *Enable flexible division of labour.* A simple role system should enable case workers to choose and work on tasks as well as trigger events they deem necessary, going beyond a restricted execute role.

REQ-13 *Make access to data flexible.* Case workers should be able to view and edit data regardless of current tasks.

How these requirements are supported and fulfilled by the framework and its architecture is described in the next chapter.

4   Architecture of the Framework

In this section we present our framework architecture and reference how it fulfills the requirements of the previous chapter. The architecture is viewed from two perspectives. A high-level perspective shows the structure of the framework itself and how it is embedded in an application server. A design-level view shows a domain model covering classes representing CMMN elements used to build case models. Also discussed are basic services provided and used by the framework.

Currently, planning elements (such as classes *PlanItemDefinition* or *PlanFragment*) are not supported by the framework and have to be transformed into elements making use of available elements (e.g. turning a discretionary item into a manually started item).

4.1   A CM Application Built with the Framework

The components of a CM application built with the framework is shown in Fig. 2. The application is run in a JEE7 application server and consists of five central layers (1–5). Access to the application is provided either directly via an integrated Java-based frontend framework or via Representational State Transfer (REST) interfaces (6). Both use stateless services provided by the framework (2). The services are described in Sect. 4.3.

External systems can interact with the application the same way as web clients, i.e. via the REST interfaces provided. These are a common approach to realize access to required web-services. The basic architecture fulfills REQ-1. These and additional components are described in this section.

The standard [20] is used to implement the framework. It fulfills REQ-2 by enabling portability among application servers and avoiding vendor-specific annotations. Furthermore, JEE7-conform application server and database management system supported by JPA can be used to run CM applications built with the framework, avoiding vendor lock-in.

The different components of the layered structure cover almost all requirements:

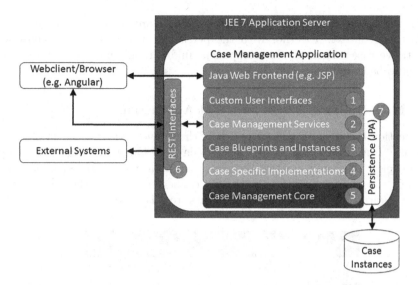

Fig. 2. Layers of the application in the JEE server environment

1. ***Custom User Interfaces*** grant case workers access to the application and include overviews of case instances, views for current tasks and data. Interfaces use services (2) to interact with the application. Together with these, REQ-3 and 4 are fulfilled by providing graphical user interfaces to case workers to access case information and data.
2. ***Case Management Services*** provide functions for central elements and access to case blueprints and instances (3). Using the services, REQ-10, 12 and 13 are fulfilled by providing access to instance data and a selection of available tasks to suitable roles.
3. ***Case Blueprints and Instances*** build on the core classes and individual implementations (4), fulfilling REQ-7 by providing case instance structures with CMMN based building blocks.
4. ***Case Specific Implementations*** include reusable custom class specialisations and implementations generated during runtime, e.g. for logic of *ProcessTask* classes or *Rules* for *decorators*. Context for the reusable components is provided by referenced element instances. These fulfill REQ-8 and 11 by implementing behaviour based on CMMN execution semantics.
5. ***Case Management Core*** elements are used by all other layers. Core classes capture the CMMN execution semantics, which fulfills REQ-1 and 7 by providing a systematic base structure. Requirement 4 is fulfilled by enabling associations of roles to tasks. Like the previous item, the core also fulfill REQ-8, 9 and 11.
6. The ***REST interfaces*** build on internal services, fulfilling requirement 5 by providing acccess to the CM application for external clients and systems.
7. ***Persistence*** is based on JPA. Persistence contexts are used by the stateless services of (2). They cover *Create, Update and Delete* (CRUD) operations, e.g.

to persist case blueprints and to retrieve and manipulate resulting instances in their context. Together with the persistent storage used to persist and retrieve case instances, this component fulfills requirement 9. Data storage has to be compatible with JPA.

4.2   Main Building Blocks for a CM Application

This section highlights the class models of the previously discussed layers **Case Specific Implementations** (4) (light green) and **Case Management Core** (5) (purple). Discussed are the *Element*, role, data and *Sentry* structures provided by the framework and how they are connected to each other. These structures constitute the layer **Case Blueprints and Instances** (3).

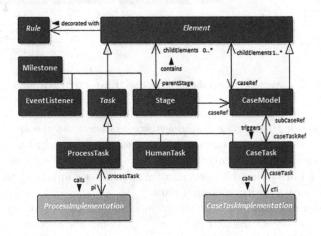

Fig. 3. Base CM elements (Color figure online)

Figure 3 shows classes representing CMMN elements. Central is the abstract class *Element* from which specialisations inherit base attributes. These include an id for persistence, the current state and *cmId* as a CM-related identifier. The *CaseModel* reference *caseRef* provides context to case instances and can be used in queries.

Internally, factory methods are used to instantiate custom implementations of layer (4) during runtime. The correct context is provided by the reference to *Elements* and *CaseModels*.

Classes *CaseModel* and *Stage* serve as containers for child elements like *Task*, *MileStone* and *EventListener*. Specialisation of abstract class *Task* further include *ProcessTask* and *CaseTask*. Both are associated with abstract classes used for case specific implementations located in layer (4).

ProcessImplementations can be used to execute algorithms, e.g. to send an e-mail or communicate with external systems such as a BPM platform, while *CaseTaskImplementation* is used to start and as a link to a nested sub-case. For this a method is overridden.

Where permitted as per CMMN specification, *Elements* can be associated with *Rules* representing CMMN decorators. Specialisations are shown in Fig. 5. Figure 4 shows a simple role system based on classes representing *CaseWorkers* (both as case admins and regular workers) and *CaseRoles* in the context of a *CaseModel*. The structure can be used to restrict access to *Tasks*. *CaseWorkers* can claim *Tasks* which are then included in their individual task list.

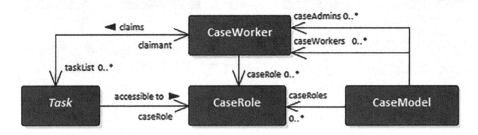

Fig. 4. Simple role system

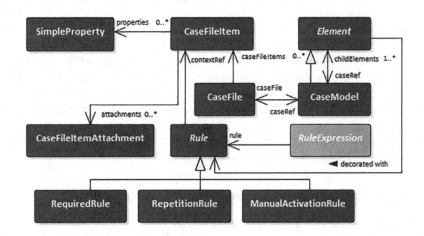

Fig. 5. Data structure

Figure 5 shows classes representing the CMMN *CaseFileItems* associated with a *CaseFile* of a *CaseModel*. *SimpleProperties* are for primitive data types to structure data, while *CaseFileItemAttachment* is added to reference unstructured data such as document files. The figure also shows how *CaseFileItems* are referenced by the three *Rule* specialisations representing CMMN decorators: *RequiredRule*, *RepetitionRule* and *ManualActivationRule*. A case specific implementation, *RuleExpression*, is used by a *Rule* to evaluate a referenced *CaseFileItem*. For this a method is overridden.

Figure 6 shows classes to build *Sentry* structures. Not displayed are specialisations *EntrySentry* and *ExitSentry*. Where permissible, these can be attached

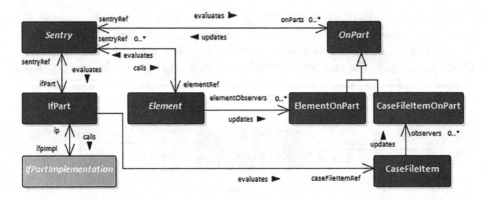

Fig. 6. Sentry structure

to *Elements*. For example, a *CaseModel* can only be associated with *ExitSentries*, while most of the other elements can have both types of *Sentry*. *ElementOnPart* and *CaseFileItemOnPart* are specialisations of abstract class *OnPart*, used to observe life cycle transitions. A *Sentry* can have an *IfPart*, which like *Rules* references a *CaseFileItem* for evaluation.

Names of the directed associations highlight the execution semantics of *Sentries*: When an *Element* or *CaseFileItem* transitions to another state, the observing *ElementOnPart* or *CaseFileItemOnPart* updates their *Sentry*. The notified *Sentry* then evaluates its (other) *OnParts* and *IfPart* if it exists. If the *Sentry* conditions are fulfilled, the *Element* it is attached to is called and its state transitioned.

How these elements are assembled into a *CaseModel* blueprint of layer **Case Blueprints and Instances** (3) is shown in Sect. 5.

4.3   Case Management Services

The framework provides basic services for CRUD operations, covering all elements making up a case, such as *CaseModel*, *Tasks* or *Milestones*. A central *CaseService* provides methods to persist, retrieve and transition (specific) case instances. A *CaseFileService* provides methods to manipulate case data, i.e. retrieve *CaseFileItems*, transition them from one state to another or to manipulate their *Properties*. Its functions can further be used for monitoring and reporting purposes.

A *TaskService* can be used to retrieve a list of (available) *(Human)Tasks* for a specific case instance, across all cases, or by role restriction. *HumanTasks* claimed by a *CaseWorker* can also be queried. The service also manages transitions of *Task* states. Repetition decorators are also managed by the service, i.e. it creates a new instance of a given task to be repeated.

Working closely with the *TaskService* is the *CaseWorkerService*, which covers operations to manage case workers, i.e. to query either for all, by id or by login credentials, as well as to create, update and delete them. A *CaseWorker* object is used in the *TaskService* to associate an available *HumanTask* to a specific worker. Other services are a *MilestoneService* and an *EventListenerService*, mainly used to retrieve information on their states or in the latter case to trigger an event.

5 Approach to Implementing Case Management Applications

In this section, we show how the elements of a CMMN model are manually translated to code using the framework. Our aim is not to parse and execute a CMMN mark up with all the intricate details and abstraction levels of the specification in a generic environment. We want to enable developers to systematically build and implement case management applications on top of light-weight core building blocks including basic services.

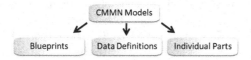

Fig. 7. CMMN model to CM application with the framework

Figure 7 shows three main components resulting from CMMN models: case model blueprints, data definitions and individual parts. These make up the layers 1–4 described in Sect. 4.1 and build on core classes located in layer 5.

```
111 HumanTask assembleSpecifications = new HumanTask("aS", "Assemble Specifications",
112          createTechnicalSpecificationCase);
113 RequiredRule requiredRule = new RequiredRule("assembleReq", specifications);
114 assembleSpecifications.setRequiredRule(requiredRule);
115
116 HumanTask reviewSpecifications = new HumanTask("rS", "Review Specifications",
117          createTechnicalSpecificationCase);
118 RepetitionRule repetitionRule = new RepetitionRule("reviewRep", specifications);
119 reviewSpecifications.setRepetitionRule(repetitionRule);

131 EntrySentry es = new EntrySentry("eSrevS", "eSrevS", reviewSpecifications);
132 new ElementOnPart(es, assembleSpecifications, StageTaskTransitions.complete.toString());
```

Fig. 8. Model to code: HumanTasks

Framework building blocks are assembled into blueprints which are used to instantiate case instances. They contain the core structures of a CMMN model. For a complete example of a blueprint structure see Fig. 11 in Sect. 6. Each element in the model is represented by a framework class.

Figure 8 shows a code example of two *HumanTasks* and the resulting code. First, two *HumanTask* are created (lines 111 and 116) and references to a *RequiredRule* (line 113f.) and *RepetitionRule* (line 118f.) are added. Their given names will be referenced later in a factory method to return the correct rule implementation.

The *EntrySentry* attached to the *HumanTask* **Review Specifications** is created and an *ElementOnPart* added in line 131f. with the transition seen in the CMMN model as the connector labelled *[complete]*.

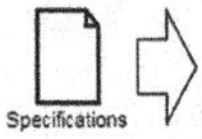

```
 98 CaseFileItem specifications = new CaseFileItem("specifications",
 99         MultiplicityEnum.ExactlyOne.toString(), "Specifications");
100 caseModel.getCaseFile().addCaseFileItem(specifications);
101 SimpleProperty required = new SimpleProperty("required", Boolean.toString(true));
102 specifications.addProperty(required);
```

Fig. 9. Model to code: CaseFileItem

Data definitions consist of *CaseFileItems* and their *Properties*. These can be included in blueprints, or added later via services. A reference to the *Case-FileItem* itself is required in the blueprint for *Sentry* references. Figure 9 shows a code example for a basic definition. The created *CaseFileItem* in line 98f. is added to the *CaseFile* of the *CaseModel* in line 100. The added Property 'required' seen in lines 101f. is used to evaluate the decorator on the *HumanTask* seen above on the left hand side in Fig. 8.

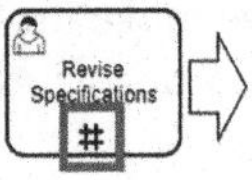

```
13  @Override
14  public boolean evaluate() {
15      CaseFileItem caseFileItem = this.rule.getContextRef();
16      boolean dataApproved = Boolean.valueOf(caseFileItem
17          .getProperty("dataApproved").getValue());
18      if(dataApproved) {
19          return false;
20      } else {
21          return true;
22      }
23  }
```

Fig. 10. Model to code: Repetition Rule

Individual parts include graphical user interfaces (GUI), *IfParts* associated with *Sentries*, decorator rules and implementations for *Process-* and *CaseTasks*. GUI are not further discussed but can use services provided by the framework to access all case instance elements, e.g. a GUI of *HumanTask* **Assemble Specification** could access the *CaseFileService* to upload specification documents.

Figure 10 shows the code needed to implement the repetition decorator on the shown *HumanTask* (note the highlight). The correct instance of the *Case-FileItem* to be evaluated is provided by the reference from line 15. Its *Property* 'dataApproved' is accessed and a boolean value is returned.

Similarly, methods of abstract classes used for *IfParts* and *Process-* and *Case-Tasks* are overridden. A factory returns the individual implementation at runtime which is then used by the framework.

Factory methods called by the framework which are used to return blueprints and individual parts need to be adjusted. Elements and their contained references, such as a *HumanTask* and its reference to its *CaseModel*, provide the correct context.

6   Case Study

Using the approach presented in the previous section, we have implemented a CM application with the framework[1], deployed on a Tomcat 8.5 TomEE PluME application server with a JTA managed MySQL data source. The presentation layer is implemented with Vaadin 8 [24] and its CDI plug-in using a *HumanTask cmId* attribute. Previously shown Fig. 1 shows the CMMN models used to assemble *CaseModel* blueprints and implement required implementations. They capture a small simplified part of a previous case study in [26].

Figure 11 shows the blueprint for the upper case *Component Release* (see Fig. 1): Here, the CMMN model is directly translated into a Java structure using framework elements such as *CaseModel*, *CaseFile*, *Stage*, *CaseTask* or *Milestone*. First, in line 68, a new *CaseModel* is created. Then in line 70ff., a *CaseFileItem* is created and added to the *CaseModel*. Given to the constructors of *Elements* is a reference to the parent: the *Stage* in line 74 is given a reference to the *CaseModel*, the *CaseTask* is constructed with a reference to that *Stage*, while the other elements such as the *Milestone* are added directly to the *CaseModel*. In the following lines, the whole structure is translated. Of note is line 90, which transitions the *CaseModel* from an initial state to state available.

Individual implementations are required for several elements. Both models are instantiated via factory method calls to get their blueprints. They are then persisted with the help of a *CaseService*. The *EventListener Cancel* can be triggered by a graphical user interface making use of an *EventListenerService* in the service layer of the framework.

Case workers instantiate the parent case from an interface which uses a *CaseService* provided by the framework. The bottom case is instantiated automatically via activation of the included *CaseTask*. It calls the blueprint of the referenced child case at runtime, which is then persisted and started. Figure 12 shows the parent and child case in a simple view listing cases. This view is provided by the framework as part of the presentation layer based on Vaadin. The referenced *CaseTask* is notified and completes once the child case automatically completes after the required *ProcessTask* is completed. While the repetition decorators require rule implementations, the *AutoComplete* decorators (both cases) do not need individual implementations. Workers have the option to claim *HumanTasks*. Figure 13 shows claimed and completed tasks of a case worker. This view is also provided by the framework as part of the presentation layer based on Vaadin.

---

[1] The framework and examples are maintained at https://github.com/fhbielefeld agpm.

```
67⊖public static CaseModel getComponentReleaseCaseModel() {
68  CaseModel model = new CaseModel("Component_Release", "Component Release");
69  model.setAutoComplete(true);
70  CaseFileItem specifications = new CaseFileItem("specifications",
71          MultiplicityEnum.OneOrMore.toString(),"Specifications");
72  model.getCaseFile().addCaseFileItem(specifications);
73
74  Stage createTechnicalSpecifications = new Stage("createTechSpecs",
75          "Create Technical Specifications", model);
76  CaseTask createSpecifications = new CaseTask("createSpecs",
77          "Create Specifications", createTechnicalSpecifications);
78  Milestone specificationsCreated = new Milestone("specificationsCreated",
79          "Specifications Created", model);
80  EventListener cancel = new EventListener("cancel", "Cancel", model);
81
82  ExitSentry exitCase = new ExitSentry("exitCase", "exitCase", model);
83  ElementOnPart cancelCase = new ElementOnPart(exitCase, cancel,
84          EventMilestoneTransitions.occur.toString());
85
86  EntrySentry entryMsSpecsCreated = new EntrySentry("enterMsSpecsCreated",
87          "entryMilestone", specificationsCreated);
88  CaseFileItemOnPart fileItemOnPart = new CaseFileItemOnPart(entryMsSpecsCreated,
89          specifications, CaseFileItemTransition.create.toString());
90  model.getContextState().create();
91
92     return model;
93 }
```

Fig. 11. Blueprint for sample process (see Fig. 1)

Fig. 12. Case list view

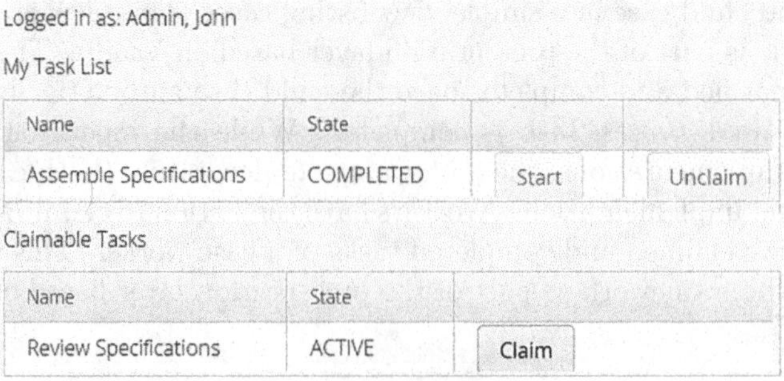

Fig. 13. Task list view

The structure of *CaseFileItem Specifications* needs to be defined in the blueprint. A *SimpleProperty* with a boolean value expressing its approval state is used to evaluate *RequiredRules* and *RepetitionRules*. The required files are either uploaded via completing *HumanTask Assemble Specifications* or via a data viewer component. Both make use of a *CaseFileService*.

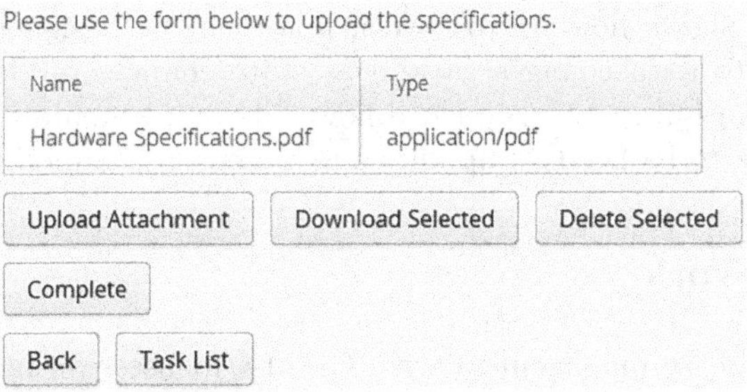

Please use the form below to upload the specifications.

Name	Type
Hardware Specifications.pdf	application/pdf

Upload Attachment   Download Selected   Delete Selected

Complete

Back   Task List

Fig. 14. Attachment upload

Figure 14 shows a PDF file was added as a *CaseFileItemAttachments* with the help of a *CaseFileService*. It is also used to update and transition the *CaseFileItem* to activate the *HumanTask* to review the specifications. *IfPart* implementations are needed to evaluate whether a revision and thus repetition of the respective *HumanTask* is necessary, or if the last update transition of the data means that it is ready for the *ProcessTask* to upload it to the parent case.

To give a sense of the amount of work needed to implement the case study application, Table 1 lists the lines of code needed to create working blueprints based on the framework. These include an implementation to execute the *CaseTask*, two *RepetitionRules*, one implementation for the execution of the *ProcessTask* and one for the *IfPart*. They do not include lines of code for user interfaces, but these range from 100 to 200 lines per Vaadin view, including controller logic.

Table 1. Lines of code required to provide working *CaseModels*

Artifact	Lines of Code
Assembled blueprints	60
Implemented decorator rules	20
Implementation of CaseTask	4
Implementation of ProcessTask	10
Implementation of IfPart	10
Total	**114**

The framework has about 4.700 lines of code organised in about 130 classes (including specialisations) in 15 packages. Graphical user interfaces were built on the framework services. These include a case list overview to instantiate and access cases, task lists to view and claim tasks, as well as GUIs for *Human-Tasks*. Provided services are used to access elements such as *CaseModel* and *Task* instances, trigger transitions of *Elements*, or to retrieve and manipulate data in *CaseFiles*.

The example shows how flexible work flows can be designed with CMMN: paths can react to data changes as well as state changes of elements without being bound to an imperative sequence flow. The implementation is supported by the framework to realize CM applications based on the models.

7   Related Work

Most research on case management with CMMN focusses on modeling aspects. Research on CMMN based implementations is available to a lesser extent.

The Darwin Wiki [13] uses a subset of CMMN in an extended wiki platform implementation to empower non-expert end-users to structure processes for knowledge work. It integrates a graphical editor using a sub-set of CMMN elements to model case work in the wiki.

A reference architecture for model-based collaborative information systems is presented in research related to the Connecare project [18]. It integrates process and data modelling in a fully model-based system enabling non-technical end-users to create case-based processes. It includes process models and a case execution engine based on an extended CMMN sub-set, though it does not further detail how.

An approach to utilize a Content Management Interoperability System (CMIS) to implement an information model based on CMMN is presented in [17]. Its focus is on using a CMIS folder as the CMMN *CaseFile* containing the case instance data, linking CMMN data concepts to existing document management systems.

Camunda [2] and flowable [11] partially support CMMN. In contrast to our framework, which focusses on creating light-weight CM applications, their main focus is on a BPMN engine and platform.

Another approach to implementing CM concepts is FLOWer, but unlike our framework it is not based in CMMN. [25] shows a simplified internal structure of FLOWer, representing artefacts such as case and activity. Activities are directly associated with data objects, forms and roles.

A fragment-based CM (fCM) case engine is presented in the Chimera project [12]. Based on the PCM approach to CM, a process is split into smaller fragments and combined with domain models, object life cycles for data objects as well as goal states. Fragments are modeled with a separate modeler ('Gryphon'). Unlike our framework, it uses elements based on a BPMN sub-set. The fragments are dynamically combined and executed depending on data states.

A data-centric business process management approach similar to but not based in CMMN is presented in research related to the PHILharmonicFlows project [16]. Users can define case-like structures and propagate ad-hoc changes of these and data to already running instances.

8   Conclusion and Future Work

Designing and implementing case management applications is a challenging and complex task. In this paper, we have presented a case management implementation framework based on Java which allows the rapid realization of case management applications based on widely available Java and open source technologies. The framework makes use of Java EE technologies and includes support for major CMMN elements. A first evaluation of our framework has demonstrated that the effort for creating a CM application using our framework is then concentrated on designing user interfaces and implementing the logic of the CMMN model. Other aspects such as case instance management and execution semantics are provided for by the framework.

Future work includes the extension of our framework to cover further elements of CMMN such as discretionary items to support an ACM approach. Other work includes the automatic generation of blueprint skeletons and specialisation stubs, the implementation of fine grained security and user management layers as well as integration with existing systems.

References

1. OMG: Business Process Model and Notation (BPMN) Specification (2011). (Version 2.0)
2. Camunda: Workflow and decision automation platform. https://camunda.com
3. Di Ciccio, C., Marrella, A., Russo, A.: Knowledge-intensive processes: characteristics, requirements and analysis of contemporary approaches. JoDS 4(1), 29–57 (2015)
4. OMG: Case Model Management and Notation (CMMN) Specification (2016). (Version 1.1)
5. Meyer, A., et al.: Implementation framework for production case management: modeling and execution. In: IEEE 18th EDOC, pp. 190–199. IEEE (2014)
6. Kurz, M., et al.: Leveraging CMMN for ACM: examining the applicability of a new OMG standard for adaptive case management. In: S-BPM ONE (2015)
7. Marin, M.A., Hauder, M., Matthes, F.: Case management: an evaluation of existing approaches for knowledge-intensive processes. In: Reichert, M., Reijers, H.A. (eds.) BPM 2015. LNBIP, vol. 256, pp. 5–16. Springer, Cham (2016). https://doi.org/10.1007/978-3-319-42887-1_1
8. Hull, R., et al.: Introducing the guard-stage-milestone approach for specifying business entity lifecycles. In: Bravetti, M., Bultan, T. (eds.) WS-FM 2010. LNCS, vol. 6551, pp. 1–24. Springer, Heidelberg (2011). https://doi.org/10.1007/978-3-642-19589-1_1

9. Tran, T.T.K., Pucher, M.J., Mendling, J., Ruhsam, C.: Setup and maintenance factors of ACM systems. In: Demey, Y.T., Panetto, H. (eds.) OTM 2013. LNCS, vol. 8186, pp. 172–177. Springer, Heidelberg (2013). https://doi.org/10.1007/978-3-642-41033-8_24
10. Fischer, L., Koulopoulos, T.: Taming the Unpredictable: Real World Adaptive Case Management: Case Studies and Practical Guidance. Future Strategies (2011)
11. Flowable: Flowable java business process engines. https://www.flowable.org/
12. BPT group. Chimera. https://bpt.hpi.uni-potsdam.de/Chimera
13. Hauder, M., Kazman, R., Matthes, F.: Empowering end-users to collaboratively structure processes for knowledge work. In: Abramowicz, W. (ed.) BIS 2015. LNBIP, vol. 208, pp. 207–219. Springer, Cham (2015). https://doi.org/10.1007/978-3-319-19027-3_17
14. Herzberg, N., Kirchner, K., Weske, M.: Modeling and monitoring variability in hospital treatments: a scenario using CMMN. In: Fournier, F., Mendling, J. (eds.) BPM 2014. LNBIP, vol. 202, pp. 3–15. Springer, Cham (2015). https://doi.org/10.1007/978-3-319-15895-2_1
15. IBM: Case management overview. https://www.ibm.com/support/knowledgecen ter/SSCTJ4_5.3.2/com.ibm.casemgmt.installing.doc/acmov000.htm. Accessed 25 Jan 2019
16. Künzle, V., Reichert, M.: PHILharmonicFlows: towards a framework for object-aware process management. J. Softw. Maint. Evol.-Res. 23, 205–244 (2011)
17. Marin, M.A., Brown, J.A.: Implementing a case management modeling and notation (CMMN) system using a content management interoperability services (CMIS) compliant repository. CoRR, abs/1504.06778 (2015)
18. Michel, F., Matthes, F.: A holistic model-based adaptive case management approach for healthcare. In: IEEE 22nd EDOCW, pp. 17–26. IEEE (2018)
19. Motahari-Nezhad, H.R., Swenson, K.D.: Adaptive case management: overview and research challenges. In: IEEE 15th CBI, pp. 264–269. IEEE (2013)
20. Oracle: Java EE 7 Technologies. https://www.oracle.com/technetwork/java/javaee/tech/index-jsp-142185.html. Accessed 25 Jan 2019
21. Schuerman, D., Schwarz, K., Williams, B.: Dynamic Case Management for Dummies: Pega. Wiley, Hoboken (2014). (Special edition)
22. Papyrus Software: Adaptive case management (2019). https://www.isis-papyrus.com/
23. Swenson, K.D., Fischer, L.: How Knowledge Workers Get Things Done: Real-World Adaptive Case Management. Future Strategies (2012)
24. Vaadin: Vaadin framework 8 (2019). https://vaadin.com/
25. van der Aalst, W.M.P., Weske, M.: Case handling: a new paradigm for business process support. Data Knowl. Eng. 53(2), 129–162 (2005)
26. Zensen, A., Küster, J.: A comparison of flexible BPMN and CMMN in practice: a case study on component release processes. In: IEEE 22nd EDOC, pp. 105–114. IEEE (2018)

Analytics

Earth Movers' Stochastic Conformance Checking

Sander J. J. Leemans[1]([✉]), Anja F. Syring[2], and Wil M. P. van der Aalst[2]

[1] Queensland University of Technology, Brisbane, Australia
s.leemans@qut.edu.au
[2] Process and Data Science (Informatik 9), RWTH Aachen University,
52056 Aachen, Germany

Abstract. Process Mining aims to support Business Process Management (BPM) by extracting information about processes from real-life process executions recorded in event logs. In particular, conformance checking aims to measure the quality of a process model by quantifying differences between the model and an event log or another model. Even though event logs provide insights into the likelihood of observed behaviour, most state-of-the-art conformance checking techniques ignore this point of view. In this paper, we propose a conformance measure that considers the stochastic characteristics of both the event log and the process model. It is based on the "earth movers' distance" and measures the effort to transform the distributions of traces of the event log into the distribution of traces of the model. We formalize this intuitive conformance metric and provide an approximation and a simplified variant. The latter two have been implemented in ProM and we evaluate them using several real-life examples.

Keywords: Stochastic process mining ·
Stochastic conformance checking · Stochastic languages ·
Stochastic Petri nets

1 Introduction

Today's information systems provide an abundance of information about activities performed by or for customers, employees, machines etc. Databases and transaction files can be converted to event logs ready for analysis. Process mining aims to provide analysts with procedures and tools to obtain insights from these recorded event logs. For instance, *process models*, which describe the process steps in the process (*activities*), which activities are to be executed and in what order activities can be executed, are used to document and prescribe processes. A process model can be obtained manually, by human analysts modelling, or automatically, by process discovery algorithms using recorded event data [1].

Both human analysts and process discovery algorithms might leave out certain behaviour of the process to make the model more readable or to capture

© Springer Nature Switzerland AG 2019
T. Hildebrandt et al. (Eds.): BPM 2019, LNBIP 360, pp. 127–143, 2019.
https://doi.org/10.1007/978-3-030-26643-1_8

only the "happy flow". In order to not limit the model to seen behaviour only, they also include other behaviour and generalise the model in this way. Therefore, before drawing conclusions from a model, it should be evaluated using a conformance checking technique. Such a technique compares a process model with an event log and highlights their differences. Using conformance checking, deviations between log and model can be unearthed, as well as differences between different versions of a business process, for instance the same process in geographic regions or different periods [8]. Furthermore, stochastic conformance checking is used to evaluate discovered process models and process discovery algorithms.

In typical real-life processes, not all parts of the processes are executed equally often: rarely executed exception handling routines or infrequent paths might be included in the model. Knowledge of the likelihood of such paths is necessary to gain insights into performance aspects of a business process, for instance to predict the remaining duration of a running trace [20] or, given a particular deadline, to estimate the probability of missing the deadline [6]. We refer to a process model that defines likelihoods for its traces as a *stochastic process model*. Such models can be automatically discovered from event logs by stochastic process discovery techniques, such as [12]. Consequently, *stochastic conformance checking* compares an event log with a stochastic process model and highlights their differences.

For instance, consider the event log $[\langle a, b \rangle^1, \langle b, a \rangle^{99}]$ and the stochastic process model expressing the stochastic language $[\langle a, b \rangle^{0.99}, \langle b, a \rangle^{0.01}]$. Even though only 2% of the log and model's stochastic language overlaps, any conformance checking technique that does not take the stochastic perspective into account will consider the log and model to have a perfect fitness and precision.

In this paper, we first propose an intuitive theoretical stochastic conformance checking measure. This measure is based on the earth movers' distance, that is, given two distributions (piles of earth), the effort to transform one pile into the other in terms of dirt that needs to be moved times the distance over which dirt has to be moved. This measure is defined for stochastic languages with possibly infinitely many traces (process models with loops may have infinitely many possible behaviours), but challenging to compute automatically for the general case. Therefore, second, we introduce an approximation and a simplification. We provide algorithms and implementations for both these last two measures, and illustrate their differences with conformance checking techniques that do not take probabilities into account.

We apply the measures to several real-life logs and automatically discovered stochastic process models to show their applicability.

In the remainder of this paper, we first explore related work in Sect. 2 and introduce concepts in Sect. 3. In Sect. 4, we introduce the three measures, after which we evaluate and analyse them in Sect. 5. Section 6 concludes the paper.

2   Related Work

Stochastic Process Formalisms. Several formalisms to describe stochastic languages have been proposed. Next to Stochastic Petri Nets (SPNs) and Generalised GSPNs which we will introduce in Sect. 3, several extensions have been proposed. For instance, Markov regenerative SPNs [18] and generally distributed transition SPNs [13] allow for the modelling of generally distributed timed events. Fluid SPNs extend SPNs with continuous fluid quantities to model physical systems [19], and controlled SPNs extend SPNs for decision support purposes [11]. For a more elaborate overview, please refer to [5], in which several types of SPN are discussed, as well as the feasibility to compute their case-duration distribution. Several of the SPN types support inhibitor arcs, but silent transitions are not supported. Typically, stochastic Petri nets are used to express and compute the temporal perspective of business processes, while we focus on the combination of the control flow and stochastic perspectives of the traces in the model. However, none of these works considers SPNs with silent transitions.

An exception is [3], in which stochastic Petri nets with silent transitions are considered. In [3], a method is proposed to, given an SPN, (a set of) initial marking(s) and a trace, compute the possible markings the SPN can be in after executing the trace, and how likely each marking is.

In Sect. 3, we introduce our formalisation of Generalised Stochastic Labelled Petri Nets (GSLPN), which differs from the variants in the mentioned papers by including silent transitions. Even though our implementation targets GSLPNs, our measures work for any stochastic process model, as long as it represents a stochastic language.

Stochastic Conformance Checking. Conformance checking on non-stochastic process models has been addressed by many techniques (e.g. token-based replay and alignments, for an overview please refer to [4]). Such techniques typically consider two directions of inclusion: fitness (behaviour of the log is included in the model) and precision (behaviour of the model is included in the log). However, these notions of language inclusion do not apply to stochastic behaviour, as a stochastic language cannot include another stochastic language (for both languages, the probabilities of traces sum to 1). Therefore, a single similarity measure is more appropriate in a stochastic setting.

In [9], standard Petri nets are enriched with frequency information (that is, each transition gets a probability corresponding to the frequency of its label in the event log), and a most-probable alignment is computed in order to find the root cause of deviations. However, this approach is not intended to cope with arbitrary stochastic languages.

Hidden Markov Models (HMMs) have been used to model stochastic processes and to check conformance, for instance in [14]. HMMs express that transitions between states and the execution of activities in states can happen with certain probabilities. In Sect. 5.4.3 of [14], fitness and precision of non-stochastic Petri nets without concurrency are computed by translating the net to an HMM assuming that all transitions are equally likely. This approach might be applicable to verify the conformance of general stochastic languages as well, as long as the language can be expressed as an HMM.

3   Preliminaries

Stochastic Languages. Let Σ be a finite alphabet of *activities* (the different steps executed in a process) and Σ^* be the set of all possible sequences (*traces*) over the alphabet Σ. A *stochastic language* is a collection of traces with attached probabilities. Formally, a stochastic language is a function $f\colon \Sigma^* \to [0,1]$ that maps each trace onto a probability such that $\sum_{t\in\Sigma^*} f(t) = 1$. For instance, $[\langle a,b,c\rangle^{\frac{2}{3}}, \langle a,c,b\rangle^{\frac{1}{3}}]$ is a stochastic language consisting of 2 traces, the first of which has a probability of $\frac{2}{3}$ and for which first an activity a was executed, followed by a b and a c. We denote the set of traces of a stochastic language M that have a nonzero probability with $\widetilde{M} = \{t \in \Sigma^* \mid M(t) > 0\}$.

Event Logs. An event log is a finite multiset of traces, which can be easily transformed into a stochastic language by normalising the trace quantities by dividing each trace's occurrences by the total number of traces. For instance, an event log consisting of 20 times $\langle a,b,c\rangle$ and 10 times $\langle b,a,c\rangle$ would have a corresponding stochastic language $[\langle a,b,c\rangle^{\frac{20}{30}}, \langle b,a,c\rangle^{\frac{10}{30}}]$. In this paper, we will use the term "event log" for the stochastic language belonging to an event log.

Earth Movers' Distance. The Earth Movers' Distance or Wasserstein distance describes the distance between two distributions [16]. In an analogy, given two piles of earth (the distributions), it expresses the effort required (in terms of quantity of earth and the horizontal distance it needs to be moved) to transform one pile into the other.

Stochastic Petri Nets. A *labelled Petri net* is a tuple (P, T, F, Σ, l), in which P is a finite set of places, T is a finite set of transitions, $F\colon (P \times T) \to (T \times P)$ is a flow relation, Σ is a finite alphabet of activities and $l\colon T \to \Sigma \cup \{\tau\}$ is a labelling function, such that $P \cap T = \emptyset$ and $\tau \notin \Sigma$. A *marking* is a multiset of places $\in P$, indicating the state of the net. A transition is *enabled* if each of its incoming places contains a token. When a transition *fires*, it changes the marking of the net by consuming and producing tokens to/from its connected places. If a transition $t \in T$ is labelled with an activity $l(t) = a \in \Sigma$, then a is executed whenever t fires. Note that multiple transitions might share a. A transition $t' \in T$ that is unlabelled ($l(t') = \tau$) is a *silent* transition: when t' fires, it may change the marking of the net but it does not correspond to the execution of an activity. A *path* through the model is a sequence of transition firings that starts in the *initial marking* and ends in a marking in which no transitions are enabled (a *deadlock*). That is, we consider each deadlock to be a final marking. The corresponding *trace* is the sequence of labelled transitions in a path.

A *Generalised Stochastic Labelled Petri Net* (GSLPN) is a tuple $(P, T, F, \Sigma, l, T_i, T_t)$ where (P, T, F, Σ, l) is a labelled Petri net, $T_i \subseteq T$ is a set of immediate transitions and $T_t \subseteq T$ is a set of timed transitions such that $T_i \cap T_t = \emptyset$. Immediate transitions $t \in T_i$ take precedence over $t' \in T_t$: timed transitions cannot fire if an immediate transition is enabled. A transition $t \in T_i$ is *immediate*, it has a *weight* $w(t)$ attached (this weight may depend on the marking) and if multiple transitions $T' \subseteq T_i$ are enabled, a transition t is chosen to fire with probability $w(t)/\sum_{t' \in T'} w(t')$. A timed transition t has an exponentially distributed waiting/enabling time, with firing rate parameter $\lambda(t)$. Due to the memory-less property of the exponential distribution, given a set of enabled timed transitions $T' \subseteq T_t$, the probability that a particular transition t will fire first is $\lambda(t)/\sum_{t' \in T'} \lambda(t')$ [12].

The probability of a trace is defined as the sum of the probabilities over all paths through the model that produce the trace. Given a trace and a path through the model that produces the trace, then the probability of the trace is the product of the probabilities of the choices made in the model along the path. The stochastic language of an GSLPN is the weighted set of all traces through the model.

For instance, Fig. 1 contains a GSLPN in which all transitions are immediate. The stochastic language of this model consists of two traces, $\langle a, b \rangle$ and $\langle a, c \rangle$. In the model, there are infinitely many paths resulting in these traces, and their probabilities are geometric series. For $\langle a, b \rangle$:

$$\frac{1}{2} + \frac{1}{2}\frac{1}{2}\frac{1}{2} + \frac{1}{2}\frac{1}{2}\frac{1}{2}\frac{1}{2}\frac{1}{2} + \cdots = \sum_{n=0}^{\infty} \frac{1}{2}(\frac{1}{4})^n = \frac{\frac{1}{2}}{1 - \frac{1}{4}} = \frac{2}{3}.$$

For $\langle a, c \rangle$:

$$\frac{1}{2}\frac{1}{2} + \frac{1}{2}\frac{1}{2}\frac{1}{2}\frac{1}{2} + \frac{1}{2}\frac{1}{2}\frac{1}{2}\frac{1}{2}\frac{1}{2}\frac{1}{2} + \cdots = \sum_{n=0}^{\infty} \frac{1}{4}(\frac{1}{4})^n = \frac{\frac{1}{4}}{1 - \frac{1}{4}} = \frac{1}{3}$$

This example illustrates that for GSLPNs, it might be challenging to establish the stochastic language.[1] To the best of our knowledge, this challenge has not been solved yet for our GSLPNs (in particular, for silent transitions) [3,15], however an in-depth discussion is outside the scope of this paper.

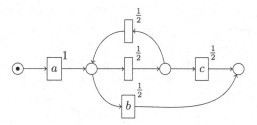

Fig. 1. Example of a generalised stochastic Petri net. All transitions are immediate.

Please note that livelocks, that is the inability to reach a final marking with nonzero probability, invalidate the stochastic language, as paths that enter the

---

[1] Notice that in case of duplicated labels, there might be exponentially, but finitely, many paths through the model for a particular trace.

livelock with a certain probability p cannot terminate, thus the total probability of the corresponding stochastic language will be at most $1 - p$ rather than 1. Therefore, in this paper, we assume that GSLPNs do not have livelocks.

The firing of transitions in GSLPNs only depends on the current state of the model: immediate transitions are randomly chosen by weight and timed transition are exponentially distributed and hence memoryless. Therefore, GSLPNs satisfy the Markov property and can be translated to Markov chains.

4   Method

In this section, we introduce our measure for stochastic conformance checking. We first introduce the theoretical measure and illustrate it with a running example. Second, we provide a method to compute the measure.

4.1   Earth Movers' Stochastic Conformance

In this section, we transform the analogy of the Earth Movers' Distance to stochastic languages: our measure expresses the cost of transforming the distribution of traces of one language into the distribution of the other language.

First, we introduce the concept of reallocation functions, which indicate how a stochastic language is transformed into another stochastic language. Second, we introduce a distance function, which expresses the cost of transforming one trace into another trace. Third, we introduce a cost function that expresses the cost of a particular reallocation function. Finally, we define the measure and we give a variant that considers unit trace distances.

Reallocation. We first introduce a function that indicates the movement of probability mass between two stochastic languages. Let L and M be stochastic languages, then a *reallocation* function $r \colon \widetilde{L} \times \widetilde{M} \to [0, 1]$ describes how L can be transformed into M. That is, $r(t, t')$ describes the probability mass of $t \in \widetilde{L}$ that should be moved to $t' \in \widetilde{M}$. The function $r(t, t)$ indicates the probability mass of $t \in \widetilde{L}$ that remains at $t \in \widetilde{M}$.

To ensure that a reallocation function properly transforms L into M the probability mass of each $t \in \widetilde{L}$ should be accounted for. Hence, the row for t should sum up to $L(t)$.

$$\forall_{t \in \widetilde{L}} \; L(t) = \sum_{t' \in \widetilde{M}} r(t, t') \tag{1}$$

Similarly, the mass of traces $t' \in M$ should be preserved:

$$\forall_{t' \in \widetilde{M}} \; M(t') = \sum_{t \in \widetilde{L}} r(t, t') \tag{2}$$

We refer to the set of all reallocation functions r that adhere to Eqs. (1) and (2) as $\mathcal{R}$ (note that $\mathcal{R}$ depends on L and M).

For instance, consider the stochastic languages $L_e = [\langle a \rangle^{\frac{1}{4}}, \langle a, a \rangle^{\frac{3}{4}}]$ and $M_e = [\langle a \rangle^{\frac{1}{2}}, \langle a, a \rangle^{\frac{1}{4}}, \langle a, a, a \rangle^{\frac{1}{8}}, \langle a, a, a, a \rangle^{\frac{1}{16}} \ldots]$. An example reallocation function r_e is:

r_e	$\langle a \rangle$	$\langle a, a \rangle$	$\langle a, a, a \rangle$	$\langle a, a, a, a \rangle$	$\langle a, a, a, a, a \rangle$	$\ldots$
$\langle a \rangle$	$\frac{1}{4}$	0	0	0	0	$\ldots$
$\langle a, a \rangle$	$\frac{1}{4}$	$\frac{1}{4}$	$\frac{1}{8}$	$\frac{1}{16}$	$\frac{1}{32}$	$\ldots$

In this tabular visualisation, Eq. (1) states that each row should sum up to the corresponding value in language L_e (e.g. the first row sums up to $\frac{1}{4}$ as $L_e(\langle a \rangle) = \frac{1}{4}$). Similarly, Eq. (2) expresses that each column should sum up to the corresponding mass in M_e.

Trace Distance. Second, a trace distance function d expresses the "distance" between traces: $d \colon \Sigma^* \times \Sigma^* \to [0, 1]$. This function is 0 if and only if two traces are equal: $d(t, t') = 0 \Leftrightarrow t = t'$. Furthermore, this function is required to be symmetrical, that is $d(t, t') = d(t', t)$.

For example, we can use the normalised string edit (Levenshtein) distance. The Levenshtein distance expresses the minimum number of edit operations required to transform a trace into another trace using the event insertion, deletion and substitution operations [10]. As this distance has an upper bound in the number of events in the longest of the two traces, it can be normalised: we choose $d_l(t, t')$ to be the Levenshtein distance divided by the maximum length of t and t'.

For instance, consider our two stochastic languages L_e and M_e again. Then, the normalised Levenshtein distance d_l is:

	$\langle a \rangle$	$\langle a, a \rangle$	$\langle a, a, a \rangle$	$\langle a, a, a, a \rangle$	$\langle a, a, a, a, a \rangle$	$\ldots$
$\langle a \rangle$	0	$\frac{1}{2}$	$\frac{2}{3}$	$\frac{3}{4}$	$\frac{4}{5}$	$\ldots$
$\langle a, a \rangle$	$\frac{1}{2}$	0	$\frac{1}{3}$	$\frac{2}{4}$	$\frac{3}{5}$	$\ldots$

Cost. Given two stochastic languages, several reallocation functions might exist. However, the Earth Movers' Distance problem aims to express the *shortest* distance between the two languages, that is, the least probability mass movement over the least distance between traces.

Therefore, the cost to transform a stochastic language L into a stochastic language M using a reallocation function r is the inner product of reallocation and distance:

$$\text{cost}(r, L, M) = r \cdot d = \sum_{t \in \widetilde{L}} \sum_{t' \in \widetilde{M}} r(t, t') d(t, t') \tag{3}$$

By construction, d returns values between 0 and 1, hence $0 \leq \text{cost}(r, L, M) \leq 1$ for any r, L and M.

For instance, considering our example with L_e, M_e and r_e again, the cost of r_e given L_e and M_e is computed as follows:

$$\text{cost}(r_e, L_e, M_e) = \frac{1}{4} \cdot 0 + 0 \cdot \frac{1}{2} + 0 \cdot \frac{2}{3} + 0 \cdot \frac{3}{4} + 0 \cdot \frac{4}{5} + \ldots$$

$$\frac{1}{4} \cdot \frac{1}{2} + \frac{1}{4} \cdot 0 + \frac{1}{8} \cdot \frac{1}{3} + \frac{1}{16} \cdot \frac{2}{4} + \frac{1}{32} \cdot \frac{3}{5} + \ldots$$

$$= \frac{1}{8} + \sum_{n=3}^{\infty} \frac{n-2}{2^n \cdot n} = \frac{13}{8} - \log 4$$

$$\approx 0.238706$$

Earth Movers' Stochastic Conformance. Finally, the Earth Movers' Stochastic Conformance (EMSC) measure is defined as the lowest cost for any reallocation function r and given L and M. To align EMSC with existing conformance checking measures, it is mirrored such that 1 indicates perfect conformance and 0 indicates the worst conformance.

$$\text{EMSC}(L, M) = 1 - \min_{r \in \mathcal{R}} \text{cost}(r, L, M) \tag{4}$$

In our running example, r_e is an optimal reallocation function, thus $\text{EMSC}(L_e, M_e) \approx 0.761294$.

Unit Distances. If we choose the trace distance function d differently, another version of EMSC appears: this function can also be chosen such that each pair of traces is either classified as equal with value 0 or as unequal with value 1. Intuitively, the Earth Movers' Distance expresses the amount of earth that has to be moved times the (possibly normalised) distance it has to be moved. The new choice for unit distances takes the distance out of the equation: the earth (traces) is either moved or not, but the distance over which it is moved is not taken into account.

Hence, in our conformance measure with unit distances, we only need to take into account how much probability mass of L is to be moved, and not where this probability mass will be put in M. This simplifies the reallocation function r considerably and removes the need for the minimisation step of Eq. (4), yielding a much simpler measure for unit trace distances uEMSC:

$$\text{uEMSC}(L, M) = 1 - \sum_{t \in \tilde{L}} \max(L(t) - M(t), 0) \tag{5}$$

Intuitively, the unit distance measure expresses the mass probability ("amount") of behaviour in L that is not supported or in surplus of the behaviour in M, without considering distances between traces.

If $\tilde{L}$ is finite, for instance if L is an event log, then this sum is finite and can be computed without constructing the full stochastic language of M explicitly; M only needs to be queried for probabilities of traces that appear in L (which might still be challenging for GSLPNs as shown in Sect. 3).

4.2 Truncated Earth Movers' Stochastic Conformance

Computing EMSC measure in the previous section poses several challenges: the stochastic language might have infinitely many traces and many reallocation functions r might apply and need to be evaluated.

To address these challenges, we introduce a derivative measure *truncated EMSC* (tEMSC) that truncates infinite languages and searches for an optimal reallocation function. In the remainder of this section, we discuss how tEMSC addresses the two challenges, we analyse the new measure and we discuss its implementation.

Handling Models with Infinite Languages. If the stochastic language M has infinitely many traces, then Eq. (3) has infinitely many terms, making EMSC challenging to compute in practice. To handle such a process model, we truncate its language to only contain a certain user-chosen mass of probability (m). We chose to do this in a breadth-first prioritised fashion, consequently shorter likely traces tend to be included before longer unlikely traces.

As a side effect, Eq. (2) does not hold for truncated stochastic languages, as the probabilities of traces in such a language do not sum up to 1. Therefore, this equation is weakened to:

$$\forall_{t' \in \widetilde{M}} \ M(t') \leq \sum_{t \in \widetilde{L}} r(t, t') \tag{6}$$

Let $\mathcal{R}'$ denote the set of all reallocation functions that adhere to Eqs. (1) and (6). Furthermore, let M_m denote a truncated stochastic language of M such that at least m of M's probability mass is included: $\sum_{t \in M_m} M_m(t) \geq m$ and $\forall_{t \in M_m} M_m(t) \leq M(t)$. Then:

$$\text{tEMSC}(L, M, m) = 1 - \min_{r \in \mathcal{R}'} \text{cost}(r, L, M_m) \tag{7}$$

Then, by construction:

Corollary 1. *Let L and M be stochastic languages. Then, for m approaching 1, tEMSC and EMSC coincide:* $\text{EMSC}(L, M) = \lim_{m \to 1} \text{tEMSC}(L, M, m)$.

In a model-model comparison context, L might have infinitely many traces as well. Then, a symmetric argument applies, but an extra requirement on the reallocation function is necessary, that is, the sum of this function should be 1.

An alternative to weakening the equation is to normalise the probability mass of the model after truncation. However, as this alters the stochastic language which is an input for the conformance calculation, we discard this option.

Another option to determine the point of truncation could be to unfold the model until we included all traces that have the same length as the longest trace of the event log. For the included traces of the model, we calculate the reallocation cost as defined before. All traces part of the excluded probability mass are longer than all traces of the event log. Based on this, we assume all excluded traces of the model to be unfitting with distance 1. Therefore, this measure gives a lower bound to EMSC. However, we leave this option for future work.

Efficient Minimisation. Given a trace distance function different from the described unit costs, the goal is to obtain an optimal reallocation function that yields minimal costs. This can be achieved by solving a linear programming problem.

Linear programming is a technique that optimises a given linear function, the objective function, with respect to given linear constraints. Based on this objective function the linear programming algorithm finds the minimal/maximal value in the region of feasible solutions defined by the constraints [17].

To find the optimal reallocation function with minimal cost, we chose Eq. (3) as our objective function. The constraints of the optimisation problem are defined by Eqs. (1) and (6). Hence, only reallocation functions that preserve the probability mass of each $t \in \widetilde{L}$ as well as the mass of $t' \in \widetilde{M}$ are valid. To complete the construction of the solution space, we define the reallocation $r(t, t')$ to be non-negative.

Considering our two stochastic languages L_e and M_e, and the distances given by the normalised Levenshtein distance function, the linear programming problem is constructed as follows:

$$\text{Minimise } r(t_1, t_1') \cdot 0 + r(t_1, t_2') \cdot \frac{1}{2} + r(t_1, t_3') \cdot \frac{2}{3} + \dots,$$

$$\text{Subject to } \frac{1}{4} \leq r(t_1, t_1') + r(t_1, t_2') + r(t_1, t_3') + \dots,$$

$$\dots$$

$$\frac{1}{2} \leq r(t_1, t_1') + r(t_2, t_1'),$$

$$\frac{1}{4} \leq r(t_1, t_2') + r(t_2, t_2'),$$

$$\dots$$

$$0 \leq r(t_1, t_1'), r(t_1, t_2'), r(t_1, t_3') \dots$$

4.3   Example and Implementation

Example. Consider the following event logs: $L_1 = [\langle a, b, c, e \rangle^{0.25}, \langle a, c, b, e \rangle^{0.25}, \langle a, d, e \rangle^{0.5}]$ and $L_2 = [\langle a, b, c, e \rangle^{0.45}, \langle a, c, b, e \rangle^{0.45}, \langle a, d, e \rangle^{0.1}]$. Furthermore, consider the GSLPNs shown in Fig. 2, and the corresponding tEMSC and uEMSC measures in Table 1. Model M_1 is a model that supports any behaviour (a *flower* model), with uniform probabilities of the individual activities and ending the trace. Intuitively, this model differs considerably from both logs L_1 and L_2, which is reflected in the low measures. Models M_2 and M_3 have the same language, but their stochastic perspective differs markedly. A conformance checking technique that is not stochastic aware would consider these models to be equivalent, resulting in equivalent measures, even though their behaviour is very different from a stochastic perspective: d is much less likely in M_3. Intuitively, M_2 is closer to L_1 than to M_3, (which is closer to L_2) and this is reflected in all our measures. Finally, from M_4 the trace $\langle a, d, e \rangle$ is missing, and, accordingly, the measures being lower for L_1 than for L_2 indicates that this trace had a higher probability in L_1.

With respect to tEMSC and uEMSC, observe that both measures are consistent in their ranking of the logs and models.

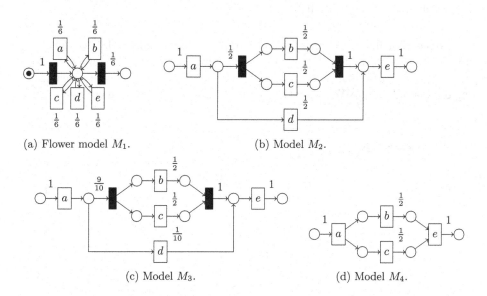

(a) Flower model M_1.

(b) Model M_2.

(c) Model M_3.

(d) Model M_4.

Fig. 2. Four example GSLPNs, in which all transitions are immediate.

Table 1. Our measures on the example logs and models.

Model	tEMSC ($m = 0.8$)		uEMSC	
	L_1	L_2	L_1	L_2
M_1	0.46	0.45	0.0010288065843622	0.0010288065843621
M_2	1	0.8	1	0.6
M_3	0.8	1.0	0.6	1.0
M_4	0.75	0.95	0.5	0.9

Implementation. Both tEMSC and uEMSC have been implemented as plug-ins of the ProM framework [7]. The plug-in of tEMSC takes an event log and an GSLPN (as returned by a stochastic process discovery technique [12]), and constructs the full stochastic language of the log, as well as the truncated stochastic language of the GSLPN, with a user-specified m. Second, a linear programming problem is constructed and solved using the LpSolve library [2]. For uEMSC, the GSLPN is not allowed to have executable loops of silent transitions (see Sect. 3). The source code of both plug-ins is available at https://svn.win.tue.nl/repos/prom/Packages/EarthMoversStochasticConformanceChecking/Trunk/.

5   Evaluation

In this section, we evaluate the newly introduced measures: the theoretical EMSC, the truncated tEMSC and the unit-distance uEMSC. First, we illustrate these on our running example. Second, we apply the measures to real-life logs to show their applicability.

Example. For illustrative purposes, we apply tEMSC and uEMSC to our running example consisting of the stochastic languages $L_e = [\langle a \rangle^{\frac{1}{4}}, \langle a, a \rangle^{\frac{3}{4}}]$ and $M_e = [\langle a \rangle^{\frac{1}{2}}, \langle a, a \rangle^{\frac{1}{4}}, \langle a, a, a \rangle^{\frac{1}{8}}, \langle a, a, a, a \rangle^{\frac{1}{16}} \ldots]$ to show the influence of truncating on a simple loop. We apply tEMSC to L_e and M_e for increasing m.

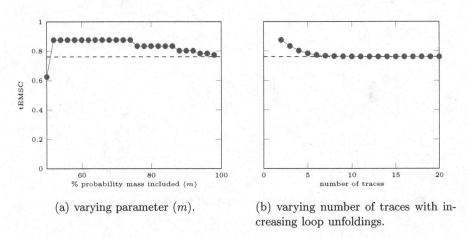

(a) varying parameter (m).

(b) varying number of traces with increasing loop unfoldings.

Fig. 3. tEMSC measured over our example L_e and M_e. The dashed lines indicate the EMSC values.

Figure 3a shows the results, as well as the theoretical EMSC value for L_e and M_e. After an initial climb when the two traces of the log are not represented by the truncated model, tEMSC temporarily stabilises at 0.875. This stable range of m indicates that the truncation includes more probability mass than m, that is, if we choose $m = 52\%$ then the truncation nevertheless includes 74% of the probability mass. Only at $m = 76\%$, another trace is included in the truncated language. After this point, the truncation includes more and more traces that are not in the event log, thus tEMSC drops and seems to approach the theoretical EMSC value shown in Sect. 4.

To illustrate the convergence to EMSC, we repeat the experiment where we manually create stochastic languages for L_e and M_e, where we unfold the loop of M_e an increasing number of times. Figure 3b shows these results. From this graph, it is clear that tEMSC quickly converges to the theoretical EMSC value with every trace and loop unfolding added. At 6 traces, which corresponds to $m = 98\%$, the difference is a negligible 0.01.

The run time for these examples was too small to warrant any conclusion.

Real-Life Event Logs. In this experiment, we evaluate the applicability of tEMSC and uEMSC to 16 publicly available real-life logs and stochastic process models. First, we apply the Stochastic Miner (SM) [12] to these logs to obtain stochastic Petri nets. As these nets contain silent transitions, they can be seen as GSLPNs (Sect. 3). Table 2 shows the logs and their complexity. Furthermore, it shows that a GSLPN was discovered for only 6 logs in the 24 BGB of RAM we had available, which illustrates the need for more research into stochastic process discovery techniques and their implementations.

Table 2. Real-life event logs used in the evaluation.

	Activities	Traces	Events	Discovery [12]	uEMSC	Rank
BPIC12	36	13087	262200	Out of memory		
BPIC15-1	398	1199	52217	Out of memory		
BPIC15-2	410	832	44354	Out of memory		
BPIC15-3	383	1409	59681	Out of memory		
BPIC15-4	356	1053	47293	Out of memory		
BPIC15-5	389	1156	59083	Out of memory		
BPIC18 Control summary	7	43808	161296	✓	0.599	1
BPIC18 Department control	6	29297	46669	✓	0.120	2
BPIC18 Entitlement application	20	15620	293245	Out of memory		
BPIC18 Geo parcel documents	16	29059	569209	Out of memory		
BPIC18 Inspection	15	5485	197717	Out of memory		
BPIC18 Parcel document	10	14750	132963	✓	$1.15 \cdot 10^{-4}$	5
BPIC18 Payment application	24	43809	984613	Out of memory		
BPIC18 Reference alignment	6	43802	128554	✓	$0.22 \cdot 10^{-2}$	3
Road Traffic Fines	11	150370	561470	✓	$2.88 \cdot 10^{-4}$	4
Sepsis	16	1050	15214	✓	$1.52 \cdot 10^{-14}$	6

Second, we apply our new measures to the logs and the six discovered GSLPNs. For tEMSC, we use various parameters m (see Eq. (7)) to study how the inclusion of mass influences the returned values. In the remainder of this section, we first discuss run times, then the results of tEMSC followed by the results of uEMSC.

Run Time. The run times of uEMSC were negligible: all measures finished within a second. For tEMSC, run times are shown in Fig. 4. Please note that y-axis is logarithmic, and that due to the inherent nondeterministic nature of tEMSC and the multithreadedness of the implementation, these measures are indicative only, especially for lower m.

Some of the values could not be obtained: for BPIC18 Parcel Document with $m = 80$, the linear programming optimisation ran out of memory, while for Sepsis with any m, the explicit creation and truncation of the stochastic language ran out of memory. For 4 out of 6 logs, computation took less than a few seconds, which was considerably less than the discovery technique, which could take hours on these logs. However, a general trend towards longer run times

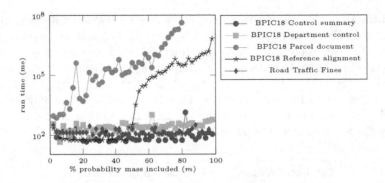

Fig. 4. Run time of tEMSC for several real-life logs.

is visible as m approaches 100%. For `BPIC18 Parcel document` and `BPIC18 Reference alignment`, computation could take up to little over an hour. A manual inspection revealed that this is caused by the size of the language described in the stochastic models: especially in models that combine concurrency with looping behaviour.

In such models, the probability mass per trace decreases and more traces are necessary to cover a certain probability mass, which makes it very challenging to obtain a high probability mass m. This is a clear limitation of the current technique, which might be addressed in future work.

*Truncated EMSC (*tEMSC*).* Second, we discuss the returned values of tEMSC, which are shown in Fig. 5 for m (that is, the minimum probability mass covered in the truncation step) varying from 2% to 98%.

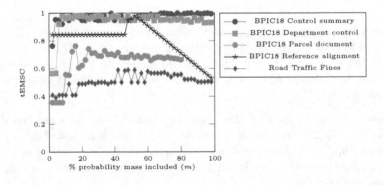

Fig. 5. tEMSC with varying mass truncation parameter (m).

Most measures show expected behaviour with increasing m: due to nondeterminism of the truncation, for lower m they vary considerably, but stabilise with m approaching 100%. An exception is `BPIC18 Reference alignment`, which increases to 0.98 for $m = 52$, after which it decreases almost linearly. A manual inspection revealed that the GSLPN contains many loops, while most of the event log's traces do not exhibit repeating activities. As m increases, more traces are added by unfolding loops and, as these new traces are not in the log, the measured tEMSC drops.

*Unit-distance EMSC (*uEMSC*).* The results for uEMSC are shown in Table 2. As identified in Sect. 3, loops of silent transitions challenge the computation of the probability of a trace in an GSLPN ($M(t)$ in Eq. (5)). A manual inspection revealed that none of the discovered models contained such loops. However, most models contained lots of concurrency, which makes the state space of the model *huge* and thus the probability of individual traces in the model is very low, leading to low uEMSC measures.

If the use case at hand involves choosing a stochastic process model that best represents a given event log, then uEMSC and tEMSC mostly agree: of the 10 possible pairs of models (out of which the model closest to the log is to be chosen), 8 times uEMSC and tEMSC agree on which model is the closest.

Reproducibility. The experiments were performed on a single machine with 3.5 GHz quadcore CPU and 24 GB RAM available for each experiment process, running fully patched Windows 7 in January 2019. The source code is available.

6   Conclusion

Recently the interest in stochastic-aware conformance checking increased within the process mining community. Despite a larger awareness about the importance of a stochastic view on the process model, to this day there are only a few conformance checking techniques that consider the stochastic characteristics of both event log and model. This paper, however, presents conformance checking measures that compare the stochastic languages of event logs and process models. In essence, this is achieved by measuring "how much probability mass that has to be moved how far" to transform one language into the other. We introduced three variations of the measure: a theoretical and two adapted versions, which are also feasible on process models with an infinite set of traces.

These adapted variants were implemented and their practical relevance was illustrated on real-life event logs. The experiments showed the influence of the probability mass parameter on the run time and the results of the measure. The evaluation showed the trade-offs between run time and memory usage, and accuracy, and it would be interesting for future work to inspect different strategies of choosing this parameter. Additionally, it would be interesting to investigate the influence of different distance functions, since the current work only compares the Levenshtein distance to a simple unit distance. However, the measure could easily be extended to use other distance functions as well.

Although EMSC simply requires two stochastic languages, its implementation starts from an event log and a stochastic Petri net. Currently, it is challenging to establish the language of a GSLPN as well as calculating the probability of a single trace in the net. Especially loops of silent transitions are shown to be problematic. Extending the technique with a new method which solves this problem will improve the reliability of the measure. Furthermore, for models with a large state space where each trace only has a low probability, the technique would benefit from a more efficient truncation implementation. Searching the model until the required probability mass has been collected is a time critical part of the measure.

In [9], a technique to calculate the most probable alignment between a model and a log is proposed, based on the probabilities of behaviour observed in the event log. For future work, it would be interesting to incorporate their technique in our reallocation function. Instead of achieving the result with the lowest cost, the algorithm would aim for the most probable reallocation.

In general, with this paper we want to stress the importance of stochastic-aware process mining and hope to inspire more discussion and contributions on this topic, for instance on the need for dual (recall/precision) measures vs. single measures or on dependencies of choices in models.

References

1. van der Aalst, W.M.P.: Process Mining - Data Science in Action. Springer, Berlin (2016). https://doi.org/10.1007/978-3-662-49851-4
2. Berkelaar, M., Eikland, K., Notebaert, P.: lp_solve 5.5. Software, 1 May 2004
3. Cabasino, M.P., Hadjicostis, C.N., Seatzu, C.: Probabilistic marking estimation in labeled Petri nets. IEEE Trans. Autom. Control 60(2), 528–533 (2015)
4. Carmona, J., van Dongen, B.F., Solti, A., Weidlich, M.: Conformance Checking - Relating Processes and Models. Springer, Cham (2018). https://doi.org/10.1007/978-3-319-99414-7
5. Ciardo, G., German, R., Lindemann, C.: A characterization of the stochastic process underlying a stochastic Petri net. IEEE TSE 20(7), 506–515 (1994)
6. Conforti, R., Fink, S., Manderscheid, J., Röglinger, M.: PRISM - a predictive risk monitoring approach for business processes. In: BPM, pp. 383–400 (2016)
7. van Dongen, B.F., de Medeiros, A.K.A., Verbeek, H.M.W., Weijters, A.J.M.M., van der Aalst, W.M.P.: The ProM framework: a new era in process mining tool support. In: Ciardo, G., Darondeau, P. (eds.) ICATPN 2005. LNCS, vol. 3536, pp. 444–454. Springer, Heidelberg (2005). https://doi.org/10.1007/11494744_25
8. van Eck, M.L., Lu, X., Leemans, S.J.J., van der Aalst, W.M.P.: PM²: a process mining project methodology. In: Zdravkovic, J., Kirikova, M., Johannesson, P. (eds.) CAiSE 2015. LNCS, vol. 9097, pp. 297–313. Springer, Cham (2015). https://doi.org/10.1007/978-3-319-19069-3_19
9. Koorneef, M., Solti, A., Leopold, H., Reijers, H.A.: Automatic root cause identification using most probable alignments. In: BPM Workshops, pp. 204–215 (2017)
10. Levenshtein, V.I.: Binary codes capable of correcting deletions, insertions, and reversals. Soviet physics doklady 10, 707–710 (1966)
11. de Meer, H., Düsterhöft, O.: Controlled stochastic Petri nets. In: SRDS, pp. 18–25 (1997)
12. Rogge-Solti, A., van der Aalst, W.M.P., Weske, M.: Discovering stochastic Petri nets with arbitrary delay distributions from event logs. In: Lohmann, N., Song, M., Wohed, P. (eds.) BPM 2013. LNBIP, vol. 171, pp. 15–27. Springer, Cham (2014). https://doi.org/10.1007/978-3-319-06257-0_2

13. Rogge-Solti, A., Weske, M.: Prediction of business process durations using non-Markovian stochastic Petri nets. Inf. Syst. **54**, 1–14 (2015)
14. Rozinat, A.: Process mining: conformance and extension. Ph.D. thesis, Eindhoven University of Technology (2010)
15. Ru, Y., Hadjicostis, C.N.: Bounds on the number of markings consistent with label observations in Petri nets. IEEE Trans. Autom. Sci. Eng. **6**(2), 334–344 (2009)
16. Rüschendorf, L.: The Wasserstein distance and approximation theorems. Probab. Theory Relat. Fields **70**(1), 117–129 (1985)
17. Sierksma, G.: Linear and Integer Optimization: Theory and Practice. Chapman & Hall/CRC, Boca Raton (2015)
18. Trivedi, K.S., Puliafito, A., Logothetis, D.: From stochastic Petri nets to Markov regenerative stochastic Petri nets. In: MASCOTS, pp. 194–198 (1995)
19. Tuffin, B., Chen, D.S., Trivedi, K.S.: Comparison of hybrid systems and fluid stochastic Petri nets. Discrete Event Dyn. Syst. **11**(1–2), 77–95 (2001)
20. Verenich, I., Dumas, M., Rosa, M.L., Maggi, F.M., Teinemaa, I.: Survey and cross-benchmark comparison of remaining time prediction methods in business process monitoring. CoRR 1805.02896 (2018)

Discovering Automatable Routines
from User Interaction Logs

Antonio Bosco[1,2], Adriano Augusto[1,3(✉)], Marlon Dumas[3], Marcello La Rosa[1],
and Giancarlo Fortino[2]

[1] University of Melbourne, Melbourne, Australia
{antonio.bosco,a.augusto,marcello.larosa}@unimelb.edu.au
[2] University of Calabria, Rende, Italy
giancarlo.fortino@unical.it
[3] University of Tartu, Tartu, Estonia
marlon.dumas@ut.ee

Abstract. The complexity and rigidity of legacy applications in large organizations engender situations where workers need to perform repetitive routines to transfer data from one application to another via their user interfaces, e.g. moving data from a spreadsheet to a Web application or vice-versa. Discovering and automating such routines can help to eliminate tedious work, reduce cycle times, and improve data quality. Advances in Robotic Process Automation (RPA) technology make it possible to automate such routines, but not to discover them in the first place. This paper presents a method to analyse user interactions in order to discover routines that are fully deterministic and thus amenable to automation. The proposed method identifies sequences of actions that are always triggered when a given activation condition holds and such that the parameters of each action can be deterministically derived from data produced by previous actions. To this end, the method combines a technique for compressing a set of sequences into an acyclic automaton, with techniques for rule mining and for discovering data transformations. An initial evaluation shows that the method can discover automatable routines from user interaction logs with acceptable execution times, particularly when there are one-to-one correspondences between parameters of an action and those of previous actions, which is the case of copy-pasting routines.

1 Introduction

The complexity and rigidity of legacy application landscapes in large organizations engender situations where workers need to perform repetitive routines to transfer data from one application to another via their user interfaces, e.g. moving data from a spreadsheet application to a Web application or vice-versa. Discovering and automating such routines can not only lead to the elimination of tedious work, but it can also reduce cycle times and improve data quality by ensuring that all data are transferred correctly.

© Springer Nature Switzerland AG 2019
T. Hildebrandt et al. (Eds.): BPM 2019, LNBIP 360, pp. 144–162, 2019.
https://doi.org/10.1007/978-3-030-26643-1_9

Robotic Process Automation (RPA) tools [1] allow us to automate such routines by recording scripts that encode sequences of interactions with Web and desktop applications, such as opening a file, selecting a field in a form or a cell in a spreadsheet, and copy-pasting data across fields/cells. While these tools allow us to automate a range of routines, they do not allow us to determine which routines to automate in the first place.

This paper presents a method to analyse User Interaction logs (UI logs) in order to discover sequences of actions (herein called *routines*) that are fully deterministic are hence automatable using RPA tools. In this context, we say that a routine is automatable if its first action is always triggered when a condition is met (the routine's *activation condition*) and the value of each parameter of each action can be computed from the values of parameters of previous actions (i.e. all actions are deterministic).

The proposed method takes as input a UI log consisting of a set of user interaction sessions (herein called *routine traces*). Each routine trace consists of a sequence of interactions (herein called actions for short). Each action has a type (e.g. select, copy, paste, etc.) and a set of parameters (e.g. the identifier of the UI element upon which the action is performed, and the inputs and outputs of the action). Given a UI log, our method outputs a set of routine specifications. A routine specification is a tuple consisting of an activation condition and a sequence of action specifications. An action specification, in turn, is a tuple consisting of an action type and a set of functions to compute the action's parameters from the parameters of previous actions.

The method starts by compressing the UI log into a Deterministic Acyclic Finite State Automaton (DAFSA). It then applies an algorithm to decompose biconnected graphs (of which a DAFSA is an exemplar) into Single-Entry Single-Exit (SESE) regions. Some of these regions correspond to sequences of actions. For each such sequence, the method checks if every action is deterministic. If so, it tries to discover an activation condition using a rule mining technique. If, on the other hand, an action in the middle of the sequence is not deterministic, the sequence is split into subsequences for which the method tries to discover activation conditions separately. A routine specification is generated for each (sub)sequence for which an activation condition is found.

The paper reports on a synthetic evaluation aimed at testing if the method can extract automatable routines with acceptable execution times.

The rest of the paper is structured as follows. Section 2 introduces a running example. Next, Sect. 3 describes our proposed approach, while Sect. 4 discusses its evaluation. Finally, Sect. 5 discusses related work while Sect. 6 summarizes the contributions and outlines directions for future work.

2   Running Example

Below, we introduce a real-life scenario to illustrate our approach. The example is inspired by work performed by the Service Improvement Team at the University of Melbourne, which applies RPA to automate various student-facing processes such as student admission. We specifically consider the task of updating the student residential data, manually performed by a university officer. Students' data is stored both in a student management system (accessible via Web interface) and on local Excel files for backup purposes. We assume that the university's student admission office is not interested in recording on its backup files the residential address of international students.

Table 1. Routine trace from the UIL, domestic student scenario.

	Action type	Action parameters			
		Param-1	Param-2	Param-3	Param-4
1	Click button	Target:Web	Label: STUDENTS		
2	Fill the text field	Target:Web	Label: ID Student	Value: 010234	
3	Press key	Target:Web	Label: ENTER		
4	Click button in row	Target:Web	Label: Update	ID Row: 010234	
5	Fill the text field	Target:Web	Label: Address	Value: 19 Parkville St, Burnley VIC 3121	
6	Fill the text field	Target:Web	Label: Country	Value: Australia	
7	Open file	Target:Excel	Name: 010234	Path: C:/Students/Australia/	Extension: .xls
8	Copy (Ctrl+C)	Target:Web	From: Address	Value: 19 Parkville St	
9	Paste (Ctrl+V)	Target:Excel	Row: 5	Column: A	Value: 19 Parkville St
10	Save file (Ctrl+S)	Target:Excel			
11	Click button	Target:Web	Label: Confirm Backup		

Table 1 shows an extract of an UI log of this task in the format generated by a *UI logger* we have developed.[1] The extract in Table 1 captures the sequence of actions performed by one employee to complete the task for a domestic student. The employee is already logged into the student management system, and she starts the task by clicking on the *students* button on the Web interface. Then, she enters the student ID in the *ID student* text field and presses the *enter* key to confirm. The Web interface displays a list of students, including the one searched (see Fig. 1). Next to each student entry, two options are available: *update* and *open*. The employee clicks on the *update* button, since she intends to update the residential data of the student. A new window opens with the student details, including the residential data, i.e. address and country. The employee types the new address and the country and, in the case of a domestic student (i.e. country is Australia), she opens the corresponding backup Excel file to copy part of the address (the street only) from the Web interface to the Excel file. She then saves

---

[1] Available at: https://github.com/apromore/RPA_UILogger.

the file changes and confirms the update on the Web interface by clicking button *confirm backup.*

In the case of an international student (Table 2), the update of the residential address on the student management system follows the same sequence of actions, but no backup is required. Thus, after the update on the Web interface, the employee clicks on the *No Backup* button, a dialogue box pops up to double check the selection, she clicks on the *ok* button, and finally clicks on the "Confirm" button to apply the update.

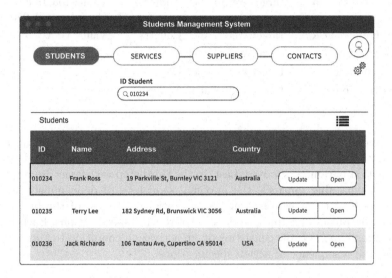

Fig. 1. Web interface of the student management system.

This example shows that a given task (e.g. updating the student residential data) may be performed via different sequences of actions (routines). In this case, there is one routine for the domestic students and another for international ones. The automation of a task requires one to identify the boundaries of each routine from the UI log, and within these routines to determine which actions are deterministic and can thus be automated.

3   Approach

In this section, we give a detailed description of the three steps composing our approach (see Fig. 2). Given as input a UI log, as first step, we parse the UI log into a DAFSA, and we extract from this latter the *flat-polygons* (the candidate automatable routines), which represent actions sequences of different length. In the second step, each of the candidate automatable routines is analysed by checking whether each of its actions is deterministic or not, i.e. the action could be executed in a systematic way by an RPA script (e.g. a software bot). The output

Table 2. Routine trace from the UIL, international student scenario.

	Action type	Action parameters		
		Param-1	Param-2	Param-3
1	Click button	Target:Web	Label: STUDENTS	
2	Fill the text field	Target:Web	Label: ID Student	Value: 010236
3	Press Key	Target:Web	Label: ENTER	
4	Click button in Row	Target:Web	Label: Update	ID Row: 010236
5	Fill the text field	Target:Web	Label: Address	Value: 106 Tantau Ave, Cupertino CA 95014
6	Fill the text field	Target:Web	Label: Country	Value: USA
7	Click button	Target:Web	Label: No Backup	
8	Click button	Target:Web	Label: Ok	
9	Click button	Target:Web	Label: Confirm	

of the second step is a set of action specifications. An action specification is a tuple consisting of: an action; and a set of functions to automatically determine all the action parameters values. Last, in the third step we extract from the candidate automatable routines, the maximal sequences of deterministic actions, and for each of them we discover the activation condition of the first action. The final output of our approach is a set of routine specifications. Each routine specification being a tuple consisting of: an activation condition to automate the routine; and a sequence of action specifications.

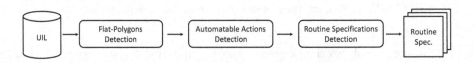

Fig. 2. Approach overview.

3.1 Definitions

The proposed approach takes as input UI logs that record multiple executions of a routine by one or several users. A UI log consists of routine traces, each one corresponding to the recording of one execution of the routine by a user. Each routine trace consists of actions that a given user performs sequentially using one or more applications (e.g. a browser and a spreadsheet application). These concepts are formalized below.

Definition 1 [Action]. *An action a is defined as a = (τ, P, V, ϕ), where: τ is the action type (e.g. click button, open file, press key, etc.); P is the set of action parameters (e.g. button name, file name, key name, etc.); V is the set of* *the values assigned to the action parameters; $\phi : P \to V$ is the function matching each action parameter to its value. Given two actions $a_1 = (\tau_1, P_1, V_1, \phi_1)$ and $a_2 = (\tau_2, P_2, V_2, \phi_2)$, a_1 and a_2 are equal if and only if (iff) they are of the same type and have the same set of parameters, i.e. $a_1 = a_2 \iff \tau_1 = \tau_2 \wedge P_1 = P_2$. Note that, two actions having the same set of parameters but with different assigned values are still considered equal.*

Definition 2 *[Routine Trace and User Interaction Log].* *A Routine Trace ρ is a sequence of actions $\rho = \langle a_1, a_2, \ldots, a_n \rangle$, we define the operator $\in$ for routine traces such that given an action $\hat{a}$, $\hat{a} \in \rho$ iff $\exists i \in [1, n] \mid a_i = \hat{a}$. Also, we refer to $a_i \in \rho = \langle a_1, a_2, \ldots, a_n \rangle, 0 < i < n$ as $\rho[i]$. Given two routine traces $\rho_1 = \langle a_1, a_2, \ldots, a_n \rangle$ and $\rho_2 = \langle \hat{a}_1, \hat{a}_2, \ldots, \hat{a}_n \rangle$, ρ_1 and ρ_2 are equal iff $\forall i \in [1, n] a_i = \hat{a}_i$. An User Interaction Log (UI log) $\mathscr{L}$ is a multiset of routine traces.*

In the proposed approach, we will sometimes reason in terms of fragments of a routine, in particular prefixes and suffixes as defined below.

Definition 3 *[Routine Trace Prefix and Suffix].* *Given a routine trace $\rho = \langle a_1, a_2, \ldots, a_n \rangle$ we define its ith prefix as $\rho^{\to i} = \langle a_1, a_2, \ldots, a_i \rangle \wedge 1 < i < n$; and its ith suffix as $\rho^{i \to} = \langle a_i, a_{i+1}, \ldots, a_n \rangle \wedge 1 < i < n$. Note that, prefixes and suffixes of routine traces (and sub-traces of routine traces) are routine traces themselves.*

Given a UI log, the goal of our proposed approach is to discover automatable routines, i.e. sets of routine traces having the parameters' values of each of their actions derivable from previous actions parameters' values.

Definition 4 *[Deterministic Action].* *Given an action $\hat{a} = (\tau, P, V, \phi)$, a UI log $\mathscr{L}$, and the set of the routine traces $R = \{\rho \in \mathscr{L} \mid \rho^{\to i}[i] = \hat{a} \vee \rho^{j \to}[1] = \hat{a}\}$, $\hat{a}$ is deterministic in $\mathscr{L}$, iff $\forall \hat{p} \in P$ one of the following holds: (i) $\phi(\hat{p})$ is constant, $\forall (\rho, a_x) \mid \rho \in R \wedge a_x = (\tau, P, V_x, \phi_x) \in \rho \wedge \hat{a} = a_x \Rightarrow \phi_x(\hat{p}) = \phi(\hat{p});$ (ii) $\phi(\hat{p})$ depends on the parameters values of the actions preceding $\hat{a}$ in the routine trace, formally, $\forall (\rho, a_x) \mid \rho \in R \wedge a_x = (\tau, P, V_x, \phi_x) \in \rho \wedge \hat{a} = a_x \Rightarrow \phi(\hat{p}) = \omega(V_1, V_2, \ldots, V_{x-1})$, where $V_i, i \in [1, x-1]$ is the set of parameters values of the action $\rho[i]$. Function ω is called a dependency function, or dependency for short.[2]*

Definition 5 *[Automatable Routine (Trace)].* *A Routine Trace $\rho = \langle a_1, a_2, \ldots, a_n \rangle$ is automatable iff $\forall a_i = (\tau_i, P_i, V_i, \phi_i), i \in [1, n]$, a_i is deterministic.*

Beyond identifying automatable routine, we seek to produce specifications thereof for their automation. Hence, we define an automatable routine specification as follows.

---

[2] In general, ω can be any function.

Definition 6 *[(Automatable) Routine Specification]. A routine specification is a tuple* $(C, \langle AS \rangle)$ *where* C *is an activation condition and* $\langle AS \rangle$ *is a sequence of action specifications. An* activation condition *is a Boolean function over a set of actions, which can be evaluated at any point in a routine of a UI log, and such that when this condition is true, an instance of the routine specification is observed. An* action specification *is a tuple* (a, Ω) *such that* a *is an action* (τ, P, V, ϕ), *and* Ω *is a set of parameter mappings. A* parameter mapping *is a tuple* $(\hat{p}, \omega)$ *such that* $\hat{p} \in P$, *and* ω *is a dependency function which computes the value of* $\hat{p}$ *from the previous actions parameters' values.*

3.2   Flat-Polygons Detection

The first step of our approach consists of detecting the flat-polygons, to do so, we execute Algorithm 1. Given as input a UI log, first, we build its DAFSA (line 1).

The DAFSA of a UI log is an acyclic automaton obtained by prefix-compressing and suffix-compressing the routine traces in the UI log. In other words, if multiple traces share the same prefix, this prefix is represented as one single sequence of states in the DAFSA, and conversely, if multiple traces share the same suffix, this suffix is represented as one single sequence of states. For details on how a DAFSA can be constructed from a set of routine traces, we refer to [13]. We observe that the DAFSA is a lossless representation of the UI log (it does not add nor remove any behavior), where each path of the DAFSA captures a different routine trace of the UI log, and each edge of a path captures an action of the routine trace, meaning that the set of all the paths in the DAFSA is exactly equal to the UI log. The prefix and suffix compression as well as the lossless feature of the DAFSA are the reasons why we chose it to represent the UI log. Indeed, capturing the UI log behavior in a lossless manner is necessary to detect and analyse the deterministic behavior recorded in the UI log. While the prefix and suffix compression allow us to easily identify each routine trace variant and where the variants start or end. Indeed, each decision point of the DAFSA (i.e. a DAFSA state with multiple outgoing edges) matches a routine variant starting point.

Once we generate the DAFSA, we compute its RPST [12] (line 2).

The RPST of a DAFSA is a *tree* where: the *nodes* are the single-entry single-exit (SESE) regions of the DAFSA; and the *edges* of the tree denote the containment relations between the SESE regions. Specifically, the children of a SESE region in the tree are the SESE regions that it directly contains. Regions at the same level of the tree represent a sequence of SESE regions in the DAFSA. Each SESE region represented by a set of DAFSA edges, depending on how these edges are related, a SESE region can be of one of four types. A *trivial* region consists of a single DAFSA edge (i.e. a routine trace action). A *polygon* is a sequence of regions (e.g. a sequence of trivial regions). A *bond* is a region where all the child regions share two common DAFSA states, one being the entry state and the other being the exit state of the bond. Any other region is a *rigid*.

The RPST allows us to detect straightforward the flat-polygons. A flat-polygon is a polygon region where all its children are trivial regions. It fol-

---

Algorithm 1. Flat-Polygon Detection

---

  input   : UIL $\mathscr{L}$

1 DAFSA $D \leftarrow$ generateDAFSA($\mathscr{L}$);
2 RPST $R \leftarrow$ generateRPST(D);
3 Set $F \leftarrow \varnothing$;

4 **for** $b \in getBonds(R)$ **do** $F \leftarrow$ extractTrivialChildren(R, b);
5 **for** $r \in getRigids(R)$ **do** $F \leftarrow$ extractTrivialChildren(R, r);
6 **for** $p \in getPolygons(R)$ **do** $F \leftarrow$ extractFlatPolygons(b);

7 **return** F;

---

lows that a flat-polygon represents a sequence of actions (DAFSA edges), these sequences are the candidate automatable routines.[3] Therefore, for each bond b in the RPST we extract the trivials that are direct children of b (see line 4), and we add each of these trivial children to the set of flat-polygons. We do the same for each rigid r in the RPST (see line 5). Whilst, for each polygon (p, see line 6) in the RPST we extract the flat-polygons as follows. If *all* the children of p are trivials, p is a single flat-polygon, otherwise, we split p into sub-polygons that have either: only rigids and/or bonds children (i.e. sub-polygons being sequences of rigids and/or bonds); or only trivial children (i.e. sub-polygons being sequences of trivials), these latter are the flat-polygons. Figure 3 shows the output of Algorithm 1 when the input UI log contains recordings of the two routine variants described in the running example. Note that we have multiple recordings of the two variants (i.e. UI log does not contain only the two routine traces pictured in Tables 1 and 2). The graph captured in the figure is the DAFSA of the UI log, whilst in blue, orange, green, and red the four flat-polygons detected and extracted.

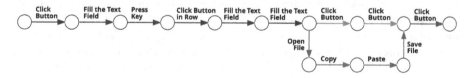

Fig. 3. Flat-polygon detection in the working example.

3.3   Automatable Actions Detection

The second step of our approach focus on the discovery of the deterministic actions, i.e. the actions that can be automated. To do so, we execute Algorithm 2. This latter receives as input the DAFSA D and the set of flat-polygons detected in the previous step (F). For each flat-polygon f in the set F, we analyse each of its actions ($a = (\tau, P, V, \phi)$), see line 3 and 4. We retrieve all the values assigned to the parameters of the action a (for all the instances of a), and we create a map Π

---

[3] Note that, a single action (a single DAFSA edge) is the simplest candidate automatable routine.

that associate to each parameter $p \in P$ all the values that the parameter assume in all the different routine traces containing the action a, $(p, \hat{V})$ (see line 6).[4] We need to collect all the values that each parameter p can assume because a is deterministic iff it is always possible to determine systematically its parameters values, this means that we must be able to compute the parameters values using constant or deterministic functions. The goal of Algorithm 2 is to identify, for each action of each flat-polygon, one function per action's parameter that allow us to deterministically compute the parameter value. A constant function is a function that assigns to a parameter always the same value, i.e. the parameter value is constant in all the instances of a. Whilst, for deterministic function, we mean a function that assigns the value to a parameter depending on the parameter values of actions executed before a. In particular, the deterministic functions we can discover are either based on *data transformation* or *substitution mappings*. We iterate on all the elements in Π, $(p, \hat{V})$, to identify these functions. First, we determine if the value of the parameter p is a constant function (see line 9), by checking if all values in $\hat{V}$ are equal.

Every time we discover a function for determining a parameter values, we add it to the list of the functions associated to the action (Ω) and we eliminate the parameter from Π (see lines 11–12, 18–19, 28–29). Doing so, if after analyzing each parameter of a, Π is empty, it means that we found for each parameter at least one function that computes its values deterministically, and therefore a is deterministic. In such case, we add a and the set of the functions to determine its parameters' values (i.e. the action specification of a) to the map α, the output of Algorithm 2.

In our running example, an action with a constant parameter is *Press Key* (see Tables 1 and 2, action #3), where regardless of the routine trace the value of the parameter *Label* is always *ENTER*. On the other hand, if the parameter p is not a constant, we check if each of its values in $\hat{V}$ is function of the previously executed actions' parameters' values. To do this, we collect all the routine traces prefixes ending with the action a (line 14). Each of these routine traces prefixes represents a routine trace variant that led to the execution of the action a, for each of this variants we try to discover the data transformation functions that allow us to compute the values of p. To discover these functions (line 15), we rely on the following two methods: (i) we look for simple value-to-value dependencies (i.e. when the value of p always matches the value of a parameter of a previously executed action); (ii) we apply Foofah, a data transfomation-by-example technique [10]; With the first method we can discover only value-to-value dependencies (the most common dependencies in RPA), hence, the deterministic function assigning a value to p would simply return the value of the matching action parameter. The second method, instead, can be used when no value-to-value dependencies are found, since Foofah can discover more complex data transformations. Foofah takes two series of data values, one called input and one called output. The input data series is the array containing all the values of one parameter of an action executed before a, whilst,

---

[4] Note that $\hat{V}$ is a list of values and not a set, i.e. it can contain duplicates.

Algorithm 2. Automatable Actions Detection

input   : DAFSA D, Flat-Polygons F

```
1  Map α ← ∅;
2  Set R ← ∅;
3  for f ∈ F do
4      for a ∈ f do
5          Boolean transformation ← TRUE;
6          Map Π ← getParametersValues(a);
7          Set Ω ← ∅;
8          for (p, V̂) ∈ Π do
9              Function ω ← identifyAndGetConstantFunction(p,V̂);
10             if r ≠ null then
11                 Π ← Π \ (p, V̂);
12                 Ω ← Ω ∪ ω;
13             else
14                 for ρ→i ∈ getPrefixRoutines(a,D) do
15                     R ← discoverDataTransformationFunctions(p, V̂, ρ→i);
16                     if R = ∅ then transformation← FALSE;
17                 if transformation then
18                     Π ← Π \ (p, V̂);
19                     Ω ← Ω ∪ R;
20                 else
21                     Matrix X ← ∅;
22                     for ρ→i ∈ getPrefixRoutines(a, D) do
23                         add rows generateMatrixRows(getActions(ρ→i), V̂) to X;
24                     Set R ← findRipperRules(X, 1.0);
25                     Decimal support ← 0;
26                     for r ∈ R do support ← support + getSupport(r);
27                     if support = 1.0 then
28                         Π ← Π \ (p, V̂);
29                         Ω ← Ω ∪ R;

30         if Π = ∅ then
31             α ← α∪ (a, Ω);

32 return α;
```

the output data series is $\hat{V}$. Foofah tries to detect functions that can determine the output from the input. Consequently, Foofah allow us to discover data transformations where: each element in the output is equal to an element in the input; or equal to a sub-string thereof; or equal to a concatenation of multiple elements in the input. Despite Foofah allow us to discover more complex data transformation functions, and detect more deterministic actions, its performance greatly affects the performance of our entire approach because: (i) it does not scale well for large input data series; and (ii) we need to use Foofah for each parameter of a and each parameter of each action executed before a (until we find a data transformation function). To partially address Foofah scalability limitation, we perform a random sampling on the input and output data series, and input these latter to Foofah. Indeed, if Foofah cannot identify a data transformation over the subsets of the complete input and output data series, it means that it could not identify any data transformation also for the complete input and output data series.

As an example of the benefits brought by Foofah, in our running example, for the routine in Table 1, the action #8 that copies only part of the student address to the Excel file can be detected as deterministic only by using Foofah.

Finally, if no data transformation functions are found for the parameter p, we try to discover functions based on substitution mappings (line 21 to 26). A function based on substitution mappings is a function that maps a set of values to another (different) set of values. For instance, a substitution mapping function can assign the value v_x to p every time that a parameter value of another action (executed before a) is equal to v_y.

To find these substitution mapping functions we use Ripper [4], an implementation of a propositional rule learner. Ripper takes as input a matrix where each row is an array of *data values* and a *label*. In our case, the data values in each row corresponds to all the values of all the parameters of all the actions executed before a in a given routine trace, whilst the label in each row matches the value of p (in the given routine trace). Ripper analyses the matrix and returns the set of rules, of which we retain those having confidence 1.0 (set R, see line 24), since such rules are the only that can be considered deterministic. Each of this rules allow us to deterministically assign a value to the parameter p during a given routine execution (captured in one or more rows of the matrix). Therefore, if all the rules with confidence 1.0 can cover all the routine executions captured as the matrix rows, it means that we can use this set of rules to deterministically assign a value to the parameter p in all the possible routine executions. To verify that the set of rules with confidence 1.0 cover all the routine executions captured in the matrix rows, we check that the sum of the rules' supports is equal to 1.0 (see lines 26 and 27).

In our running example, a substitution mapping function is detected to determine the value of the parameter *Label* in the last *Click Button* action (in both Tables 1 and 2). Indeed, every time the country value is equal to *Australia* the *Label* value of the last *Click Button* action is *confirm backup*. Whilst, every time the country value is not *Australia* the *Label* value of the last *Click Button* action is *confirm*.

3.4   Routine Specifications Detection

In the last step we identify, by applying Algorithm 3, the set of routine specifications. The algorithm receives as input the DAFSA D, the set of flat-polygons F, and the map α containing the actions specifications that we discovered with Algorithm 2. First, we extract the automatable routines from the flat-polygons, line 2. Precisely, given each flat-polygon, we check that each of its actions is deterministic (i.e. the action is in α), and we extract all the sequences of deterministic actions not interrupted by non-deterministic ones. This means that from a flat-polygon we can extract one or more automatable routines. Once detected the automatable routines (set S), we have to discover for each of them the activation condition. An activation condition of an automatable routine in S is the trigger that determines the start of the routine execution. It consists of: (i) a triggering action, i.e. an action executed just before the first action of the automatable

---

Algorithm 3. Routine Specifications Detection   ·

    input   : Map α, Flat-Polygons F, DAFSA D

```
 1  Set Ξ ← ∅;
 2  Set S ← extractAutomatableFlatPolygons(α, F);
 3  for ρ ∈ S do
 4  │   Set B ← ∅;
 5  │   Action a₁ ← ρ[1];
 6  │   for ρ→ⁱ ∈ getPrefixRoutines(a₁,D) do  B ← B∪ getActions(ρ→ⁱ⁻¹);
 7  │   Matrix X ← ∅;
 8  │   for ρ̂ ∈ getPaths(D) do
 9  │   │   Set Q ← getActions(ρ̂) ∩B;
10  │   │   if Q ≠ ∅ then
11  │   │   │   Boolean label ← ρ ⊆ ρ̂;
12  │   │   └   add rows generateMatrixRows(Q, label) to X;
13  │   Set R ← findJripRules(X, 1.0);
14  │   Boolean activationConditionFound ← TRUE;
15  │   for r̂ ∈ getRowsWithLabel(X, TRUE) do
16  │   └   if ¬ existRule(R, r̂) then  activationConditionFound ← FALSE;
17  │   if activationConditionFound then
18  │   │   Ξ ← Ξ∪ (ρ, R);
19  │   │
20  │   else
21  │   │   if hasTrivialCondition(ρ) then
22  │   │   │   R ← generateTrivialCondition(ρ);
23  │   │   │   Ξ ← Ξ∪ (ρ, R);
24  │   │   else
25  │   │   │   if |ρ| > 1 then
26  │   │   │   │   R ← generateTrivialCondition(ρ²→);
27  │   │   └   └   Ξ ← Ξ∪ (ρ²→, R);
28  return Ξ;
```

---

routine; and (ii) a boolean condition that must be valid at the completion of the triggering action. The boolean condition can be: (i) a function of parameter values of the actions executed before the triggering action (this included); or (ii) can be based only on the completion of the triggering action, i.e. the boolean condition is the completion of the triggering action. We call the first case a *data-based activation condition*, whilst the second case a *trivial activation condition*.

We discover each automatable routine (ρ) activation condition as follows. First, we collect all the routine traces prefixes ending with the action a_1, being this latter the first action of the automatable routine (line 5). Then, for each of these prefixes, we collect its actions into the set B (excluding a_1, see line 6), B will contain all the actions that can be executed before a_1. We note that the execution of any of these routine traces prefixes is not a sufficient condition for the execution of a_1 (and the automatable routine thereof), because any of these prefixes could also lead to the execution of an action different than a_1 (i.e. a_1 is an outgoing edge of a decision point of the DAFSA). To analyse for what prefixes actions' parameters values a_1 is executed or not, we can use again Ripper, similarly to Algorithm 2. In this case the Ripper input is a matrix X where: each row is the array containing all the parameters values of all the actions executed before a_1 in a given routine trace, plus a boolean label set to *true* if the automatable routine was executed within the given routine trace, *false* otherwise. To build the matrix X, we collect all the paths of the DAFSA[5] (see line 8) and, for each of them $(\hat{\rho})$, we take the actions that are contained in B (see set Q, line 9). If the set Q is not empty, we add to the matrix X a row containing all the parameters values of all the actions in Q and we set the boolean label in the row as *true*, if the automatable routine ρ is contained in $\hat{\rho}$, otherwise as *false* (see lines 11 and 12).

Once we built the matrix X, we input it to Ripper, which outputs the set of rules, of which we retain those having confidence 1.0 (set R, see line 13) Each of these rules tell us which parameters values of all the actions in B triggered the automatable routine. Then, we filter the matrix X retaining only the rows having the boolean label *true* and, similarly to Algorithm 2, we check that the rules discovered with Ripper can cover all the rows of this filtered matrix, line 16. To perform this check, we cannot rely on the support of the rules, like in Algorithm 2, because we are not interested in covering all the rows of the matrix X, but only those having the boolean label *true* (i.e. we used the full matrix to discover the rules, but we want the rules to cover only the cases when the automatable routine ρ is executed). If the rules discovered by Ripper cover all the rows of X capturing an execution of ρ (line 17), it means that we found a data-based activation condition, which is represented by the set of rules discovered with Ripper.[6] When we find an activation condition for an automatable routine, we create the routine specification (the tuple: automatable routine and activation condition) and we add it to the set of routine specifications Ξ, see line 19. On the other hand, if a data-based activation condition cannot be found with Ripper (line 22), we check if exists a trivial activation condition, i.e. if ρ can be triggered by the completion of an action preceding a_1. If we find such a trivial activation condition, we use it to create the routine specification for ρ (lines 22 and 23). Otherwise, if ρ is not a single-action routine (line 25), we can always use a_1 to generate a trivial activation condition for the subroutine $\rho^{2\rightarrow}$. Indeed being ρ a sequence of deterministic actions, it follows that after the execution of its first action (a_1), the successive actions will be executed as well.

4   Evaluation

We conducted an experiment to assess the ability of our approach to correctly discover all the automatable (sub)routines recorded in a set of UI logs. To this end, we generated nine artificial UI logs (from Coloured Petri Nets, CPNs), each log containing a different number of automatable (sub)routines of varying complexity,[7] and used the characteristics of these routines (position and actions within) as a ground truth for our experiment. These logs emulate a controlled

---

[5] We remind that a path of the DAFSA corresponds to a routine trace in the UI log.

[6] Note that, the set of rules take into account also the triggering action.

[7] The CPNs and the logs used for our evaluation are available at https://doi.org/10. 6084/m9.figshare.7850918.v1.

recording environment where user tasks are performed without noise (i.e. events that capture actions that are irrelevant to the task are not present).

For testing purposes, we packaged the implementation of our method as a Java command-line application,[8] and executed it on a PC with Intel Core i7-6600U@2.60 GHz CPU with 12 GB RAM, running Ubuntu 16.04 LTS (64-bit) with 8 GB RAM and JVM 11 (4 GB RAM). We conducted all tests using both methods to discover data-transformation functions, i.e. with and without Foofah.

4.1   Test Case Generation

We generated nine artificial UI logs by simulating nine CPNs designed with the CPN Tool [9]. The first six CPNs are simple and represent common real-life scenarios, capturing clear routines with a specific goal. CPN1 represents the following sequence of actions: the user opens a random file, opens a specific webpage, logs in with his credentials (assumed to be always the same and correct), awaits the response from the server, and then begins to copy data from the Web page to the opened file. All these actions are automatable, except the first one, since the input file is chosen randomly. CPN2 is a variant of CPN1 and it includes the handling of an error during the login action, i.e. the user enters wrong credentials. CPN3 is a further extension of the task captured in CPN2, where the user unsuccessfully repeats the login actions until they decide to quit the procedure. CPN4 is derived from CPN1, but in this task we injected non-deterministic actions among the automatable ones. We used these non-deterministic actions to perturb the fully automatable task in CPN1. These are: random button clicks (i.e. non data-driven clicks) and user data inputs (i.e. the login credentials are different in each routine trace). The number of non-deterministic actions increases in CPN5, where only 16% of the total actions is automatable, and no automatable routines are captured, i.e. there are no two consecutive automatable actions. CPN6 is the running example presented in Sect. 2, which has a balanced number of automatable and non-automatable actions.

The last three CPNs have the highest complexity and the routines they represent are not easily interpretable (e.g. they do not follow a routine goal). We decided to include these latter CPNs to evaluate the robustness of our approach in the case of complex scenarios. CPN7 has only 25% of automatable actions, which are intertwined with the non-automatable actions. CPN8 extends CPN7 by adding a long deterministic subroutine to be executed when a specific condition is met. CPN9 is the most complex case; it merges CPN5 and CPN6 and captures the situation in which a user first inputs several data and then decides the sequence of actions to perform based on the input data. Using this latter CPN we can assess our approach's ability to discover activation rules based on long-dependencies, i.e. dependencies between non sequential actions.

Table 3 reports the structural complexity of each CPN in terms of size, control flow complexity, and structuredness, as well as the statistics of the UI logs

---

[8] The software is available at http://apromore.org/platform/tools, *Automatable Routines Discoverer* package. The source code can be found at https://github.com/apromore/RPA_AutomatableRoutinesDiscoverer.

simulated from the CPNs. Precisely, we reported for each UI log the number of routine variants, the number of distinct routine traces, and the number of actions recorded.

Table 3. Characteristics of the CPNs and UI Logs.

ID	CPN			UI logs		
	Size	CFC	Struct.	#Routines	#Traces	#Actions
1	15	0	1.00	1	100	1400
2	20	2	1.00	2	1000	14804
3	20	3	1.00	6	1000	14583
4	18	4	1.00	1	100	1400
5	18	11	1.00	12	1000	8775
6	16	2	1.00	2	1000	9998
7	24	10	0.29	7	1500	14950
8	39	11	0.56	8	1500	17582
9	65	24	0.83	18	2000	28358

4.2 Results

Table 4 shows the results of our experiment. For each log generated from the corresponding CPN, we report the number of distinct automatable actions (AA) recorded, their percentage on the total number of distinct actions, the length of the longest sequence of AA (i.e. longest autom. routine length, LRL), the number of distinct AA discovered and the LRL discovered with our approach (both with and without Foofah). We note that w/ Foofah, our approach could discover all the AA as well as the LRL, except for log L3, which we discuss later. However, when disabling Foofah, we could not detect some data transformations (i.e. the non value-to-value transformations), which were necessary to identify some AA. Consequently, our approach w/o Foofah could not identify all the AA (as per L3, L6, L8, and L9), and in some cases (see L4, L6 and L9) the LRLs were shorter than those discovered w/ Foofah. Despite the Foofah variant of our approach allow us to detect more complex data transformations, and therefore discover more AA and longest autom. routines, Foofah brings an overhead in the execution time. Indeed, when enabling Foofah, our approach becomes up to 50x slower (see L9). This is due to how Foofah looks up for data transformations: its heuristics search explores a large number of possible combinations of data transformations before declaring that no data transformation occurred.

The only log where the approach failed to discover two routines is log L3, which contains loops. The reason for this limitation is that the approach takes a DAFSA (which is acyclic) as a starting point and the DAFSA does not capture the notion of loop. Repetition in a DAFSA shows up in the form of a polygon

Table 4. Experimental results on the artficially generated UI logs.

UIL	Recorded AA(#)	Recorded AA(%)	Recorded LRL	AA discovered (w/ Foofah)	AA discovered (w/o Foofah)	LRL discovered (w/ Foofah)	LRL discovered (w/o Foofah)	Exec. time (s) (w/ Foofah)	Exec. time (s) (w/o Foofah)
L1	13	92.9	13	13	13	13	13	3.0	2.8
L2	16	84.2	5	16	16	5	5	61.1	5.2
L3	17	94.4	7	15	15	6	6	224.8	7.6
L4	8	47.1	4	8	7	4	3	79.2	2.7
L5	2	16.7	1	2	2	1	1	49.6	9.4
L6	9	69.2	4	9	8	4	2	80.0	4.5
L7	4	25.0	2	4	4	2	2	279.7	8.2
L8	18	60.0	13	18	15	13	13	282.2	10.9
L9	24	68.6	5	24	22	4	3	935.2	19.3

followed by a branching in which one of the polygons is identical to the previous polygon (and this pattern can be repeated multiple times). Our approach does detect that the polygon inside the body of the loop is a routine and it discovers its activation condition, but it fails to discover the activation condition of the task that follows the exit point of the loop, particularly when this condition depends on the number of times the loop has been executed.

5   Related Work

There are resemblances between the discovery of automatable routines from UI logs and Automated Process Discovery (APD) from event logs [3], stemming from the fact that the inputs have similar structure. However, APD focuses on discovering control-flow models, without data flow. Some approaches enhance discovered process models with branching conditions [5], but we are not aware of any work on APD that discovers data transformations. Moreover, APD approaches are not suitable for automable routine discovery because they generalize the behavior in the log, i.e. they produce models that generate traces not observed in the log. Most of these algorithms also under-approximate the log's behavior, i.e. they intentionally produce models that do not perfectly fit the log. In contrast, we seek to discover only sequences of actions that have been observed in a UI log. Automating a sequence of actions that has never been observed is risky, because the cost of letting a software bot do something that should not be done is high (e.g. it may introduce spurious data into a system leading to costly mistakes and time-consuming corrective actions). This is the reason why our approach uses a lossless representation of event logs (DAFSAs) to discover candidate automatable routines, as opposed to an automatically discovered process model.

Recent related work in APD deals with the problem of discovering process models from low-level event logs [8]. In this context, a low-level event log is one where each event refers to a step of a task in a process, e.g. a task "Contact customer" is captured via several events corresponding to steps "Retrieve the customer's contact details from CRM system", "call customer", etc. The goal of [8] and related studies is to group low-level events into coarser-grained ones in order to discover conceptual models, as opposed to automatable routines as we do.

Another related family of techniques is sequence mining [2], where the goal is to discover frequent patterns in collections of sequences. In contrast to sequence mining, automatable routine discovery takes as input UI logs consisting of actions with parameters, as opposed to sequences of symbols.

Automatable routine discovery is also related to Web usage mining and UI log mining. Web usage mining seeks to discover and analyze sequential patterns in Web data, such as Web server logs capturing user interactions with Web apps [16]. Analyzing such data can help to optimize the functionality of Web apps, optimize their navigation structure, and provide personalized content to users [11].

Research proposals in UI log mining, such as TaskTracer and TaskPredictor, have tackled the problem of analyzing UI logs generated by desktop applications to identify the current task performed by a user and to detect switches between tasks [7,14,15]. These techniques do not deal with the problem of identifying routines. More closely related is the work in [6], which proposes a technique to extract frequent sequences of actions from noisy UI logs. However, this technique does not extract automatable routines because it does not discover data transformations nor activation conditions.

6   Conclusion

We introduced an approach to discover automatable routines from UI logs. Each such automatable routine is characterized by an activation condition and a sequence of action specifications, each associated with data transformations that compute an action's parameters values from those of previously observed actions. An experimental evaluation allowed us to validate that the approach can re-discover repetitive routines synthetically injected in a UI log. The evaluation also allowed us to compare two alternative methods to discover data transformations: one based on an algorithm to discover one-to-one dependencies and the other based on Foofah, which can discover complex (1-to-N, N-to-1, and N-to-N) dependencies. The evaluation showed that the one-to-one dependency discovery method scales up to relatively large logs, while putting into evidence scalability limitations of the Foofah alternative.

The evaluation highlighted one of the limitations of the approach: its inability to discover activation conditions of routines that immediately follow the exit point of a loop, particularly when this condition depends on the number of executions of the loop. This is due to the fact that DAFSAs do not capture loops. A possible approach to address this limitation is to extend the DAFSA with explicit loops. This would require us to detect those loops in the first place, for example using tandem repeat detection algorithms. This extension should be such that it allows us to reason about the number of occurrences of a loop in a routine trace, so that this information can be used when discovering the activation condition (i.e. the condition under which the loop is exited).

Another limitation of the approach is its inability to deal with noise. The presence of an event in a routine trace that is not related to the trace (e.g. an error) leads to the polygon capturing that routine being broken down into two flat-polygons. As a result, our approach will either discover only sub-routines of an otherwise automatable routine, or not discover the routine at all. Some of the noise could be removed via simplification rules that take advantage of the idempotence of some UI operations (e.g. if a user mistakenly clicks twice the same cell consecutively, this can be reduced to a single click). But more generally, it may require tailor-made noise filtering techniques. Designing such techniques is an avenue for future work.

In the embodiment of the approach presented in this paper, we used RIPPER to discover activation conditions and Foofah to discover complex data transformations. Experimenting with alternative methods to discover activation conditions and data transformations is another direction for future work.

Acknowledgements. This research is funded by the Australian Research Council (DP180102839), the Estonian Research Council (IUT20-55), and the European Research Council (project "PIX").

References

1. van der Aalst, W.M.P., Bichler, M., Heinzl, A.: Robotic process automation. Bus. Inf. Syst. Eng. **60**, 269 (2018). https://doi.org/10.1007/s12599-018-0542-4
2. Agrawal, R., Srikant, R.: Mining sequential patterns. In: ICDE. IEEE (1995)
3. Augusto, A., et al.: Automated discovery of process models from event logs: review and benchmark. TKDE **31**(4), 686–705 (2019)
4. Cohen, W.W.: Fast effective rule induction. In: ICML. Morgan Kaufmann (1995)
5. de Leoni, M., Dumas, M., García-Bañuelos, L.: Discovering branching conditions from business process execution logs. In: Cortellessa, V., Varró, D. (eds.) FASE 2013. LNCS, vol. 7793, pp. 114–129. Springer, Heidelberg (2013). https://doi.org/10.1007/978-3-642-37057-1_9
6. Dev, H., Liu, Z.: Identifying frequent user tasks from application logs. In: Proceedings of the 22nd International Conference on Intelligent User Interfaces (IUI 2017), pp. 263–273. ACM, New York (2017). https://doi.org/10.1145/3025171.3025184
7. Dragunov, A.N., Dietterich, T.G., Johnsrude, K., McLaughlin, M.R., Li, L., Herlocker, J.L.: Tasktracer: a desktop environment to support multi-tasking knowledge workers. In: IUI. ACM (2005)
8. Fazzinga, B., Flesca, S., Furfaro, F., Pontieri, L.: Process discovery from low-level event logs. In: Krogstie, J., Reijers, H.A. (eds.) CAiSE 2018. LNCS, vol. 10816, pp. 257–273. Springer, Cham (2018). https://doi.org/10.1007/978-3-319-91563-0_16
9. Jensen, K., Kristensen, L.M., Wells, L.: Coloured Petri nets and CPN tools for modelling and validation of concurrent systems. STTT **9**(3–4), 213–254 (2007)

10. Jin, Z., Anderson, M.R., Cafarella, M.J., Jagadish, H.V.: Foofah: transforming data by example. In: SIGMOD. ACM (2017)
11. Liu, B.: Web usage mining. In: Web Data Mining. Data-Centric Systems and Applications. Springer, Heidelberg (2007). https://doi.org/10.1007/978-3-642-19460-3_12
12. Polyvyanyy, A., Vanhatalo, J., Völzer, H.: Simplified computation and generalization of the refined process structure tree. In: Bravetti, M., Bultan, T. (eds.) WS-FM 2010. LNCS, vol. 6551, pp. 25–41. Springer, Heidelberg (2011). https://doi.org/10.1007/978-3-642-19589-1_2
13. Reißner, D., Conforti, R., Dumas, M., La Rosa, M., Armas-Cervantes, A.: Scalable conformance checking of business processes. In: Panetto, H., et al. (eds.) OTM 2017. LNCS, vol. 10573, pp. 607–627. Springer, Cham (2017). https://doi.org/10.1007/978-3-319-69462-7_38
14. Shen, J., et al.: Detecting and correcting user activity switches: algorithms and interfaces. In: IUI. ACM (2009)
15. Shen, J., Li, L., Dietterich, T.G.: Real-time detection of task switches of desktop users. In: IJCAI (2007)
16. Srivastava, J., Cooley, R., Deshpande, M., Tan, P.: Web usage mining: discovery and applications of usage patterns from web data. SIGKDD Explor. Newsl. **1**(2), 12–23 (2000)

Grounding Process Data Analytics in Domain Knowledge: A Mixed-Method Approach to Identifying Best Practice

Moe Thandar Wynn$^{(\boxtimes)}$, Suriadi Suriadi, Rebekah Eden, Erik Poppe, Anastasiia Pika, Robert Andrews, and Arthur H. M. ter Hofstede

Queensland University of Technology, Brisbane, Australia
{m.wynn,s.suriadi,rg.eden,erik.poppe,
a.pika,r.andrews,a.terhofstede}@qut.edu.au

Abstract. The often used notion of 'best practice' can be hard to nail down, especially when a process involves multiple stakeholders with conflicting interests, as is common in healthcare, banking, and insurance domains. This exploratory paper presents a novel method that leverages both domain knowledge and historical precedence as recorded in IT systems to derive relevant dimensions, measures and behaviours representing best practice. To test our approach, we explored best practice in the area of injury compensation claims management involving multiple stakeholders. We evidence that best practice can be identified by semi-structured interviews with stakeholders (a qualitative method) allowing their perspectives to guide the application of various forms of analytics on historical data (a quantitative method). This led to the identification of four best practice dimensions: process fairness, process quality, process cost, and process timeliness and their respective measures, which are then used to assess the performance of compensation claim cases (i.e., 'which claims are the best performing cases?'). By analysing the process behaviours of those cases through historical data together with additional stakeholder input, we propose to identify potential best practice behaviours.

Keywords: Best practice · Process data analytics · Interviews · Insurance claims

1 Introduction

Within the business process management community, the use of process data analytics (process mining [1]) to extract valuable insights about as-is processes for the purposes of facilitating well-targeted process improvement initiatives has seen process mining being increasingly adopted in today's organisations. One approach to improving one's processes is to adapt best practices that are relevant for the domain the processes belong (e.g., APQC Knowledge Base). In this paper, we regard 'best practices' as behaviours that lead to favourable process

© Springer Nature Switzerland AG 2019
T. Hildebrandt et al. (Eds.): BPM 2019, LNBIP 360, pp. 163–179, 2019.
https://doi.org/10.1007/978-3-030-26643-1_10

outcomes for one or more parties involved in the process. While best practices may be well-studied in certain domains (such as reference models in supply chain management), in other domains, there are no known best practice standards, apart from domain experts' knowledge. In cases such as insurance claims processing where multiple stakeholders possess conflicting interests (e.g., insurers, claimants, law firms, and health providers), identifying best practice can be very challenging. In this situation, one needs to first unpack the meaning of a 'best practice' process from the perspectives of different stakeholders.

From a process mining point of view, certain 'tangible performance measures of processes' (e.g., time and cost) can be readily extracted from historical data [26]. However, other equally important dimensions that are intangible and/or domain-specific, are difficult to analyse using out-of-the-box process mining techniques. For example, the notion of an 'equitable outcome' in a process involving participants with unequal power distribution (e.g., an insurer deciding payouts for an unemployed claimant), perceived or otherwise, can be translated into various context-dependent variables that are often not readily measurable. Even for 'quantifiable' variables, competing interests may complicate the way one identifies best practice processes. Understanding the implications of these insights for best practice cannot be garnered purely from quantitative data rather qualitative data is also required.

The main contribution of this paper is a mixed-method approach to identify best practices, combining qualitative (i.e., interviews with stakeholders) and quantitative analysis (i.e., historical data), we propose a three-stage approach that derives best practice for processes of a specific domain by:

- defining and extracting dimensions, variables and measures of best practice from stakeholder interviews,
- automatically discovering cases satisfying the measures related to the notion of 'best practice' from historical data, and
- extracting insights from historical data into how those cases are executed with a view of deriving best practices in collaboration with stakeholders.

To demonstrate our approach, we use a case study of injury compensation claims processing. This paper is organised as follows. Section 2 discusses prior work on best practices and the need for mixed-method approaches. Section 3 details the proposed approach. Section 4 applies the approach to injury compensation claims processing. Section 5 concludes the paper.

2   Related Work

Best practices – defined as behaviours that lead to favourable process outcomes for one or more parties involved in the process – are proffered to be associated with improvements in process efficiency, effectiveness, and quality [8,22]. As such, they are a fundamental topic across a number of domains, including: business process management. To explore best practices, scholars have predominantly used either quantitative (e.g., [18]) or qualitative (e.g., [22]) techniques in isolation.

From a quantitative perspective, historical data has been used to identify deviations from best practices, however the degree to which a data-driven approach is used for deriving best practices differ. For instance, in [29], the authors developed a tool to quantitatively measure performance indicators specified by the IT department, but no comprehensive qualitative methods and analysis techniques were used. Alternatively, others derive best practices from literature and only quantitatively assess there impacts on organizational performance [9]. Several process mining methodologies have been proposed [1,12], where the need for stakeholder's input is recognised. However, these methodologies are generic, targeting any process mining analysis. To the best of our knowledge, there is no data-driven approach for deriving aspects of best practice from process data utilising process mining.

From a qualitative perspective, literature reviews, interviews, and focus groups have been used to identify best practice. In [28], the authors review business process redesign literature to ascertain the best practice. In [22], the authors extended this work using a descriptive survey to identify 10 best practices, mapping them to the Devil's Quadrangle of cost, quality, time, and flexibility. However, these studies identify best practices they are based on perceptual data, objective data of how the process unfolds are not examined.

While each perspective provides insights into best practice, a more complete understanding can be ascertained through adopting a mixed method design [31]. In regards to process mining, quantitative techniques are limited to data contained within event logs, with typically only a surface level understanding of what the attributes represent. These data sets often contain data quality issues leading to spurious findings [6]. A qualitative approach can help mitigate these limitations through (i) providing contextual understanding which is necessary when exploring domain specific settings (e.g., best practice claims management); (ii) providing deeper insights into representations underlying data sets (e.g., identifying best practice dimensions).

Despite the importance of mixed methods for establishing completeness, developing inferences, and corroborating findings, there is a dearth of mixed methods research both in the process mining community and the broader Information Systems field [3,31]. This is also reflected within our specific context of injury compensation claims processing, where existing studies have been built based on either qualitative or quantitative approaches. In [17], the authors consider fundamental aspects of best practices for accident compensation claims management and state that "This paper reviewed the lessons learned from best practice case management research" [17]. Conversely, in [5], the authors used data mining and process mining techniques to identify key factors contributing to delays in claims processing, but did not explore how qualitative methods can complement their studies to identify best practices. In this paper, we propose a mixed-method approach to address the identification and analysis of best practice cases and demonstrate our approach for injury compensation claims management.

3   Approach

In many domains the notion of best practice is concerned with more than just processing cases as quickly or as cheaply as possible. Therefore, we propose to use both quantitative and qualitative techniques to capture additional dimensions of best practice relevant to the target domain. We involve multiple stakeholders to derive a well-rounded view of the notion of 'best practice'. Furthermore, due to the complexity and different interpretation of what constitutes 'best practice', we stress the need to involve stakeholders in the validation of findings. Broadly, we use the term 'best practice' to describe process behaviour that leads to outcomes desired by one or more stakeholders involved in the process. Since, 'best practice' is linked to outcomes of the process, we can more specifically define 'best practice' as the behavioural difference between cases achieving these outcomes and cases not achieving these outcomes (or achieving them to a lesser degree).

We propose a three-stage approach, depicted in Fig. 1, to determine best practice for processes of a specific domain. First, we gather dimensions and variables that indicate cases with desirable outcomes and develop performance measures for these outcomes. Second, we apply these measures to historical event log data to identify best performing cases for these measures. It is critical at this stage to untangle behavioural, contextual and compositional effects impacting performance measures, so that one compares apples to apples. To achieve this, the cases that should behave similarly are clustered. Third, we analyse cases classified as best performing to determine behaviour and context factors that distinguish them from similar cases that were not.

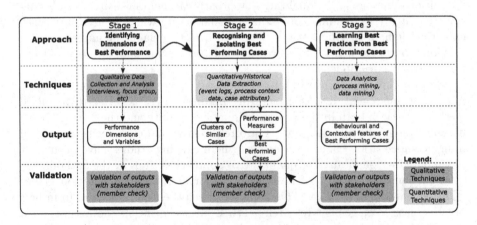

Fig. 1. A mixed-method approach to uncovering a best practice process

3.1   Stage 1: Identifying Dimensions of Best Performance

Our approach starts with forming a contextual understanding of a domain-specific 'best practice' process (e.g., injury compensation claims processing), with the goal of (1) unpacking key *dimensions* underlying best performing cases; and (2) identifying potential *variables*, observed or latent, that can be used to assess the dimensions. We detail the data collection and analysis procedures below.

Design qualitative data collection procedures. To identify domain-specific dimensions of best performing cases, qualitative data is necessary as it provides a deeper understanding of the context. There are three key considerations in designing the qualitative component: the data collection method, protocol, and sample. A variety of qualitative methods, including observations, focus groups, and interviews, have the potential to aid in identifying the dimensions. We opt for one-on-one semi-structured interviews [24] as when integrated within a mixed-method study they enable researchers to delve deeper into core concepts [15].

For the semi-structured interviews, an interview protocol is developed and structured into three sections. The first section seeks to understand how the participant is embedded within the process. The second section asks broad questions about best performing cases allowing for inductive analysis and then narrows the focus to key performance dimensions that are either present in literature or identified during previous interviews enabling constant comparative and deductive analysis [13]. The third section attempts to uncover contextual factors where the evaluation of best performing cases may ultimately differ (e.g., claimant complexity, legal representation).

With a process often involving multiple stakeholders with conflicting interests, purposeful sampling [14] is performed to ensure diverse roles and levels of seniority are accounted for. During data analysis, the sampling procedure shifts to theoretical sampling, which "is the process of data collection for generating theory whereby the analyst jointly collects, codes, and analyses his data and decides what data to collect next and where to find them, in order to develop his theory as it emerges" [16].

Identify dimensions through qualitative analysis. To analyse qualitative data, techniques from grounded theory are used. Grounded theory is an "inductive theory discovery methodology that allows the research to develop a theoretical account of the general features of the topic while simultaneously grounding the account in empirical observations of data" [23]. This technique enables novel findings to be obtained [30]. In applying grounded theory techniques, open coding is first used to identify salient concepts [13]. Next, the emergent concepts are constantly compared to each other until a refined set are defined. Subsequently, theoretical coding is performed [16] to identify *dimensions, variables, antecedents or consequences* of best practice. This process of open coding, constant comparison, and theoretical coding is performed until theoretical saturation is reached (i.e., when no new dimensions, variables or relationships are observed [16]).

Continuously validate outputs. During each stage of our mixed method approach, the findings are validated with stakeholders, primarily by performing member checking. Member checking is performed through providing results in the form of reports and presentation to both participants of the study and other stakeholders involved in the process so they can provide feedback [10]. Based on feedback, continual refinements to dimensions, variables, and measures can occur. This is essential for ensuring credibility [10]. Moreover, central to the mixed-method nature of the approach, triangulation of data from different sources (i.e., interviews, event logs) facilitates accurate definitions of the findings [31]. In addition, transitioning to theoretical sampling enables different participants not previously sampled to provide perspectives, which can be used to corroborate findings [7].

3.2   Stage 2: Recognising and Isolating Best Performing Cases

In the second stage, analysis is performed on historical data related to the process. The historical data extract could be event logs (i.e., recorded traces of activities performed by various resources - manual or automated - in the execution of the process), case attributes (e.g., the severity of injury suffered by claimants), as well as contextual data pertaining to the process (e.g., claimants demographics, changes in legislation governing insurance processes, etc). The goal of this stage is to assess the extent to which performance variables identified from Stage 1 can be measured using historical data to identify cases representing best practice. This stage consists of four steps.

Define performance measures. Firstly, we identify appropriate *measures* for each performance variable identified in Stage 1. For example, two performance variables related to the time dimension could include (i) whether or not a particular activity has been executed within a legislated time frame, and (ii) the overall duration of cases. For the first variable, the relevant measure could be the number of days or hours a particular case was late in meeting the deadline, while for the second variable, the relevant measure could include the number of days between two milestones signifying the start and end of a process (e.g. accident notification date in an injury compensation claim to settlement date).

Cluster similar cases. Cases vary from each other in terms of: behaviour, context and performance. These features also influence one another as, for example, both behaviour and context are likely to affect performance of a case. In claim processing, performance can be affected by injury severity. To remove confounding effects of cases that are (expected to be) dissimilar in context and/or behaviour, we first split cases into clusters that are expected to behave in a similar way and then classify the performance of cases within the same cluster.

Calculate and aggregate performance measures. Next, performance measures are calculated for each case. Because each dimension can have multiple variables (and hence multiple measures), we aggregate the values for all measures (pertaining to the same dimension). Different measures for a particular dimension can have different units of measurement and therefore, we normalise the values of each measure to make them comparable. Aggregation of measures per dimension can then be done by taking the statistical average or median together with configurable weights. For each case in the data, an aggregated value (normalised to lie between 0 and 1) is computed for each dimension. Then another aggregation (using a weighted average) is performed across all normalised values for each dimension such that each case is now represented by one aggregated value. Using this final value, the cases are ranked.

Identify best performing cases. Using the aggregated measure, we identify, from historical data, cases that result in desirable outcomes. As we are using measurement values to differentiate a best practice case from other cases, a threshold is needed to separate cases. That is, those cases that meet a certain threshold can be considered as best performing cases. Such a threshold could be, for example, the top 10% or 25% of cases, ranked using the aggregated measurement value. In order to achieve actionable insights, the threshold value should be configurable and derived in consultation with stakeholders.

3.3   Stage 3: Learning Best Practice from Best Performing Cases

The last stage of our approach involves the use of various analytics to extract best practice behaviours from best performing cases, i.e. what are the characteristics and behaviours of cases that make them produce the desired outcomes? To do so, various data mining and process mining techniques can be used. As discussed previously, member checking is also used to validate the findings to ensure that best practice behaviours are appropriately derived.

Conduct Process mining and Data mining. In order to identify differences in process-related behaviours, various classes of process mining techniques exist [1]. One could discover the process models for best practice cases and non-best practice cases, and compare the discovered process models to extract differences in the process behaviours [2]. One could also apply comparative process analysis tools to extract the behavioural differences between best practice and non-best practice cases visually [32]. Since cases are labeled based on their performance during Stage 2, one could also apply classic supervised machine learning techniques [19] to gain further insights into characteristics and context of cases which are classified as best practice cases or otherwise. For example, one could use classification analysis, regression analysis or contrast set learning to understand whether some case attributes are correlated with best performing cases.

4   Case Study

We applied the approach detailed in Sect. 3 in a case study involving injury compensation claims in Queensland, Australia. Processing injury compensation claims is complex, as it involves negotiations among multiple parties (e.g., claimants, insurers, law firms, and health providers). In Queensland, the injury-compensation claims scheme, known as the Compulsory Third Party (CTP) scheme, is governed by the Motor Accident Insurance (MAI) Act 1994 and is underwritten by four licensed, commercial insurers who accept applications for insurance and manage claims on behalf of policyholders. The scheme is overseen by the Motor Accident Insurance Commission (MAIC) with Nominal Defendant (ND), a statutory body, responsible for managing claims where the 'at fault' vehicle is unregistered or unidentified. Despite legislation mandating certain milestones for claims processing and providing various pathways for claims to be progressed and finalised, MAIC see significant behavioural and performance variations in CTP claims processing and variations in costs and duration of claims. The variations indicate best-practice guidelines may be needed to ensure consistent and fair outcomes. We therefore focused our data collection efforts on both MAIC and ND as MAIC ensures claims are effectively handled and ND represents an exemplar case as they typically handle more complex claims than other CTP insurers.

4.1   Identifying Dimensions of Best Performance for CTP Claims (Stage 1)

Qualitative data collection. As outlined in Sect. 3.1, semi-structured interviews were used to identify the best performing dimensions for CTP insurance claims. Using purposeful sampling, we identified two participants from the legislator (MAIC) and three from the insurer (ND) with differing roles and levels of experience. This enabled a more complete understanding of best practice claims management to emerge from the legislator and insurer viewpoints. During interviews, the influence lawyers had on claims was salient. As such, we progressed to theoretical sampling and conducted two interviews with defendant lawyers. Their insights were consistent with the legislator and insurer. The interviews were recorded, transcribed and uploaded in NVivo.

Qualitative analysis. Through open coding, constant comparison, and theoretical coding, we identified four dimensions of best performing claims: *process fairness, process quality, process costs,* and *process timeliness* (Table 1).

Foundational to these dimensions is that ND is a model litigant who needs to *"do the right thing and whatever is fair and appropriate"* (Participant 2), which means supporting legitimate claimants in such a way that the public levy for CTP insurance remains affordable. Accordingly, the *process fairness* dimension recognises the outcomes of the claims process should return legitimate claimants to the state they were in, where possible, prior to the accident by providing an appropriate settlement with timely access to rehabilitation. As such *process fairness* consists of two variables *compensation fairness* and *rehabilitation appropriateness*. It is important to note that *fair compensation* recognises all claimants are unique and should be duly compensated.

While *process fairness* is the most salient dimension and focused on the outcomes of the process, many respondents also emphasised the importance of *process quality*. *Process quality* takes into account equitable treatment of claimants

Table 1. Defining the dimensions of best practice CTP claims management

Item	Dimension	Definition
PF	Process fairness	The extent to which the outcome of a legitimate claim is perceived as fair in terms of both compensation and access to rehabilitation
PF1	Compensation fairness	The extent to which a claimant's settlement is appropriate based on injury severity and economic loss
PF2	Rehabilitation appropriateness	The extent to which a claimant receives rehabilitation in a timely manner
PQ	Process quality	The extent to which the claims management process is handled in an equitable manner with transparent communication
PQ1	Equitable treatment	The extent to which all claimants are treated equally regardless of demographic differences or legal representation
PQ2	Communication transparency	The extent to which the insurer clearly explains the claims process and outcomes to the claimant
PQ3	Liability determination	The extent to which the necessary evidence is collected to reasonably determine whether the insurer is responsible for handling the claim
PQ4	Investigation appropriateness	The extent to which the necessary evidence has been collected to justify the settlement.
PC	Process costs	The sum of the costs associated with claims handling
PT	Process timeliness	The extent to which the stages within the claims management process meet legislative requirements

and the need to transparently communicate with claimants. *Process quality* consists of four variables *equitable treatment, communication transparency, investigation inappropriateness,* and *liability determination timeliness*.

Process timeliness and *Process costs* were also considered to be core dimensions. The *Process timeliness* consists of two variables to capture whether deadlines imposed by legislation are met, and whether the overall claim (from start to finish) is handled in a timely manner. *Process costs* included all costs associated with handling the claim (e.g., legal, investigations related to liability determination and investigation appropriateness, and rehabilitation handling) excluding settlement and rehabilitation. No specific thresholds were discussed for process costs. However, participants acknowledged the need to minimise costs to ensure CTP remains feasible.

4.2   Recognising and Isolating Best Performing Cases (Stage 2)

This section elaborates on how we conducted Stage 2 of our approach using data sets provided by MAIC.

Identifying quantifiable performance measures. We analysed 31570 claims finalised between 2013 and 2018 by five CTP insurers. For each claim, there are up to 58 attributes relating to milestone dates, costs, injury, claimant and solicitor, enabling us to quantify six of the identified variables.

Two measures for *Process Timeliness* were determined. There are three legislative provisions relating to processing times of a CTP insurer: (i) assess the compliance of a claim notification and respond within 14 days, (ii) determine appropriate rehabilitation measures for the claimant within 14 days and (iii) determine liability within 6 months. As the relevant dates for notification, compliance determination, rehab decision and liability decision are present in the data set, the first measure of timeliness is computed by summing the number of days a claim takes over given thresholds (PT1). As the legislation only covers part of the CTP claims process, another time measure related to the overall duration of a claim is also computed (PT2).

The cost of a claim for an insurer can be divided into processing cost and the payout. The *Processing Costs* (PC) include the cost of investigations for the determination of liability and appropriateness of payout, the legal costs and various other operational costs for the insurer. It is computed by the sum of these cost items from the data set. A measure of fairness should capture whether the claimant was awarded the appropriate amount of money to cover economic loss, medical care and rehab required as well as general damages relating to the severity of their injuries. The amount of general damages based on injury severity is scheduled in the Civil Liability Regulation of 2014. As both injury severity (ISV) and general damages paid are available in the data set, a measure for *Compensation Fairness* (PF1) can be computed to determine the difference between paid amount and amount scheduled in the legislation. A measure for *Rehabilitation Appropriateness* (PF2) can be derived from the number of days between claim notification date and rehab decision date.

Equitable Treatment (PQ1) relates to how claimants are treated across different claims. We use a method of measuring similarity that is, for example, used in cluster analysis (e.g. [27]), which is the standardisation of measures and use of squared Euclidian distance from each cluster's median. Other variables are conceptually more complex and the three remaining *Process Quality* variables could not be operationalised with the given data set.

Calculating performance measures. We first computed the measures and then computed the normalised values of all measures for each dimension. Figure 2 illustrates the decomposition and measurement of CTP best performing dimensions. The first Process Timeliness measure (PT1) relates to legislative guidelines as described previously. The second Process Timeliness measure (PT2) is calculated as the difference of days between the notification and finalised dates to

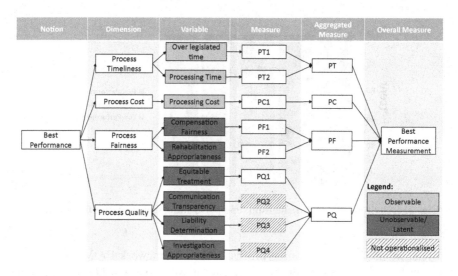

Fig. 2. Illustration of the decomposition of our best performance dimensions.

measure overall processing duration. Processing cost (PC) summarises the processing costs for a claim. The Compensation Fairness measure (PF1) is calculated as the difference between general damages scheduled for an ISV according to the Civil Liability Regulation 2014, and the actual amount paid. The Rehabilitation Fairness measure (PF2) is computed as number of days until access to rehabilitation is provided to a claimant. The most complex measure is that of Equitable Treatment (PQ1). Firstly, we computed the median across (normalised values of) PT1, PT2, PC, PF1 and PF2 for all cases. The value of PQ1 for a particular case is then the difference between the vector of measurements for this case and the vector of medians.

Identifying best performing cases. An overall ranking of claims by "goodness" of practice can be achieved by using a weighted average of the values of all dimensions. In this study we use equal weights for all dimensions, but depending on the domain, stakeholders can use the weights to prioritise certain dimensions. Using this value, cases were binned into different categories and labeled as such (i.e., the top 25% of cases as high-performing cases; the middle 50% as medium-performing and the bottom 25% as low-performing cases). Figure 3 shows how cases are distributed among five CTP insurers: Insurer 4 has the highest number of best performing cases and Insurer 3 has the smallest number.

4.3 Learning CTP Best Practice from Best Performing Cases (Stage 3)

We analysed data from one of the insurers shown in Fig. 3, the *Nominal Defendant* (ND) insurer, recorded for claims finalised between 2012 and 2018. In the

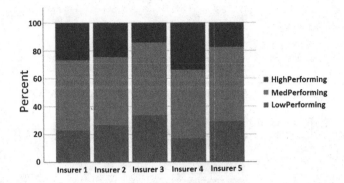

Fig. 3. Distribution of high/medium/low-performing cases among five insurers.

analysis, we used an open source process mining framework ProM[1], which provides functionality to load, filter and transform event logs and has many plugins to enable all kinds of process analysis.

As per the organisation's request, we first grouped claims based on values of several attributes which are known to affect claim performance (age, injury severity, legal representation and vehicle category). The majority of the resulting clusters only included few cases (ranging from 3 to 39); hence, they were not used for further analysis. Case performance was separately evaluated for four largest clusters which included 279, 238, 218 and 105 claims. The insurer changed the information system which supports the process in 2014 and complete process execution data is only available for claims that started after the system change. We selected cases that started after the system change and are finalised and used them to discover process differences between high-performing and low-performing cases. The selected cases included 147 high-performing cases, 176 cases with medium performance and 42 low-performing cases.

The process data contained 51 activities and all claims followed unique process paths. We applied decision tree classification (using activity executions in cases as descriptive variables) and identified activities that were associated with high-performing or low-performing cases. The event log was filtered to include four such activities (shown in Fig. 4) and also activities "Upload new claim" and "Finalised" which are performed in all cases. We then selected cases that follow five most frequent process paths for high-performing cases (117 out of 147 high-performing cases (i.e., 80%)) and cases that follow five most frequent process paths for low-performing cases (36 out of 42 low-performing cases (i.e., 86%)) and discovered a process model from the resulting event log using the "Inductive miner" [20] ProM plug-in. The selected high-performing and low-performing cases were replayed (separately) on the discovered process model using the "Multi-perspective Process Explorer" [21] ProM plug-in; the results are depicted in Fig. 4.

---

[1] www.promtools.org.

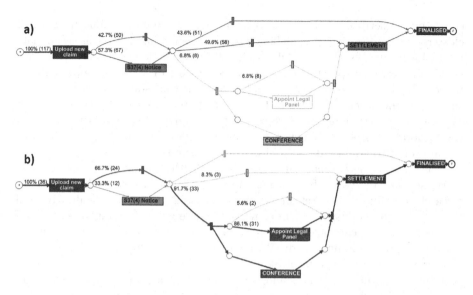

Fig. 4. Main process differences between (a) high-performing and (b) low-performing cases.

Figure 4a demonstrates that in high-performing cases activity "Appoint Legal Panel" was not performed and conference was held in only 6.8% of cases; while Fig. 4b shows that both activities were performed in most low-performing cases. We can see that settlement was performed in all low-performing cases and was not performed in 43.6% of high-performing cases; while activity "S37(4) Notice" is more frequently performed in high-performing cases (57.3%) than in low-performing cases (33.3%). These findings were presented to some Nominal Defendant stakeholders and they commented that the findings are consistent with their expectations: the conference and the appointment of the legal panel in a claim indicate a higher complexity of the claim, which is expected to have a higher cost and a longer processing time and cases without settlement are expected to be faster and cheaper. The stakeholders noted that our initial findings explain time and cost dimensions well; in future, we plan to focus on fairness and quality to determine if we can identify specific process behaviours that explain performance differences for these dimensions.

4.4 Discussion

In this section, we reflect on the exploratory case study conducted to determine the feasibility of the proposed approach. In line with requirements of many data analysis techniques, our approach requires sufficient, high-quality historical data to gain reliable insights. This requirement is particularly pertinent when the data needs to be split into multiple clusters in order to eliminate confounding effects. For instance, we were provided with claims data for all CTP claims completed by *Nominal Defendant* (ND) insurer over a seven-year period. However, in the case study described in Sect. 4.3, we chose not to analyse claims from small clusters as best performance labelling based on a small number of cases can yield unreliable results. Furthermore, if data is collected from long periods of time, one should also consider possible changes in the process or systems supporting the process. In the ND case study, we analysed claims that started after the system supporting the process was changed (as complete process execution data was not available for claims lodged before the system change). Such observations from the exploratory study conducted with the ND data set enable us to fine-tune our approach to analyse data sets from other insurers.

The use of methodological triangulation with the intent of overcoming "intrinsic bias that comes from a single-method" is well-known in the qualitative research community [11]. Yet, in the process mining community, qualitative and quantitative methods are seldom combined. There is limited emphasis on 'how' and 'when' to engage stakeholders to obtain actionable insights during various stages of a process mining study. Without clear methodological guidance, quantitative process data is susceptible to the fallibility of the data set (especially with respect of incompleteness and inaccuracy of data), potentially resulting in content validity issues. The importance of content validity cannot be understated and is purported to be the most important type of validity as it is "the extent to which a measurement reflects the specific intended domain of content." [25].

Our methodology (c.f. Fig. 1) enhances the content validity of the process mining findings through methodological triangulation. The qualitative data collection at the commencement of the study aids in improving content validity ensuring domain specific insights are leveraged. Moreover, iterative member checking improves the credibility of our findings.

While our methodology can overcome several limitations apparent with purely objective or subjective data collection, there are still some limitations present. Where possible we tried to mitigate these. For instance, there is an inherent risk in qualitative interviews that interviewees can be biased and satisficing [4]. We mitigated this risk through voluntary participation, guaranteeing confidentiality, and diverse participants who had different roles and responsibility allowing for cross comparisons. Moreover, the objective data minimised this risk through triangulation. Another limitation of this approach is it captures perceptions of what the best practice dimensions are at a point in time, yet changes over time could change the dimensions present. We argue, that when an organization is evaluating best practices they want it to be based on current policies and ideologies, and important changes over time would be raised during interviewing or member checking.

In this paper, we demonstrated the value of complementing data-driven insights gained from process mining analysis with domain knowledge extracted from stakeholders through the use of interviews, demonstrations, and member checking in the context of uncovering a best-practice CTP claims process. In regards to generalisability, the purpose of this research is not to generalise our findings to all domains, rather it is to provide an example of a mix-method approach that other scholars can employ to identify domain specific best practice.

5   Conclusions

This paper proposes a mixed-method approach to derive a best practice process by combining qualitative data collection techniques (i.e., interviews/focus groups with stakeholders) and data-driven analysis techniques (i.e., data mining and process mining techniques). The proposed approach is illustrated through an injury compensation claims management scenario where interviews with stakeholders were conducted and claims data from multiple insurers were analysed. This paper presents the results from the first iteration where we have asked multiple stakeholders to identify dimensions of interest, measures and thresholds and then conducted the analysis of the data sets provided to determine the overall feasibility of the approach. We have shown how the dimensions of best practice described by stakeholders such as *process fairness*, *process quality*, *process cost*, and *process timeliness* can be linked to measures that are computed from the data sets. Future work will focus on a detailed analysis of historical data including context data together with member checking to derive best practice guidelines across all four dimensions.

Acknowledgments. The work presented in this paper was funded by a grant from the Queensland Motor Accident Insurance Commission (MAIC).

References

1. van der Aalst, W.M.P.: Process Mining: Data Science in Action. Springer, Heidelberg (2016). https://doi.org/10.1007/978-3-662-49851-4
2. van der Aalst, W.M.P., Adriansyah, A., van Dongen, B.F.: Replaying history on process models for conformance checking and performance analysis. Data Min. Knowl. Discov. **2**(2), 182–192 (2012)
3. van der Aalst, W.M., Dustdar, S.: Process mining put into context. IEEE Internet Comput. **16**(1), 82–86 (2012)
4. Alshenqeeti, H.: Interviewing as a data collection method: a critical review. English Linguist. Res. **3**(1), 39–45 (2014)
5. Andrews, R., et al.: Exposing impediments to insurance claims processing. In: vom Brocke, J., Mendling, J. (eds.) Business Process Management Cases. MP, pp. 275–290. Springer, Cham (2018). https://doi.org/10.1007/978-3-319-58307-5_15
6. Bose, J., Mans, R., van der Aalst, W.: Wanna improve process mining results? It's high time we consider data quality issues seriously. In: CIDM, pp. 127–134. IEEE (2013)
7. Charmaz, K., Bryant, A.: Grounded theory and credibility. Qual. Res. **3**, 291–309 (2011)
8. Cho, M., Song, M., Comuzzi, M., Yoo, S.: Evaluating the effect of best practices for business process redesign: an evidence-based approach based on process mining techniques. Decis. Support Syst. **104**, 92–103 (2017)
9. Christmann, P.: Effects of "best practices" of environmental management on cost advantage: the role of complementary assets. Acad. Manag. J. **43**(4), 663–680 (2000)
10. Creswell, J.W., Miller, D.L.: Determining validity in qualitative inquiry. Theory Pract. **39**(3), 124–130 (2000)

11. Denzin, N.: Sociological Methods. McGraw-Hill, New York (1978)
12. van Eck, M.L., Lu, X., Leemans, S.J.J., van der Aalst, W.M.P.: PM^2: a process mining project methodology. In: Zdravkovic, J., Kirikova, M., Johannesson, P. (eds.) CAiSE 2015. LNCS, vol. 9097, pp. 297–313. Springer, Cham (2015). https://doi.org/10.1007/978-3-319-19069-3_19
13. Fernández, W.D., et al.: The grounded theory method and case study data in is research: issues and design. In: Information Systems Foundations Workshop: Constructing and Criticising, vol. 1, pp. 43–59 (2004)
14. Flick, U.: An Introduction to Qualitative Research. Sage Publications Limited, Thousand Oaks (2018)
15. Galletta, A.: Mastering the Semi-structured Interview and Beyond: From Research Design to Analysis and Publication. NYU Press, New York (2013)
16. Glaser, B.: Doing Grounded Theory: Issues and Discussions. Sociology Press (1998)
17. Iglesias, M., Walsh, J.: Accident compensation claims management-lessons learnt and claimant outcomes (2009)
18. Kis, I., Bachhofner, S., Di Ciccio, C., Mendling, J.: Towards a data-driven framework for measuring process performance. In: Reinhartz-Berger, I., Gulden, J., Nurcan, S., Guédria, W., Bera, P. (eds.) BPMDS/EMMSAD -2017. LNBIP, vol. 287, pp. 3–18. Springer, Cham (2017). https://doi.org/10.1007/978-3-319-59466-8_1
19. Kotsiantis, S.B.: Supervised machine learning: a review of classification techniques. In: Emerging Artificial Intelligence Applications in Computer Engineering (2007)
20. Leemans, S.J.J., Fahland, D., van der Aalst, W.M.P.: Discovering block-structured process models from event logs containing infrequent behaviour. In: Lohmann, N., Song, M., Wohed, P. (eds.) BPM 2013. LNBIP, vol. 171, pp. 66–78. Springer, Cham (2014). https://doi.org/10.1007/978-3-319-06257-0_6
21. Mannhardt, F., De Leoni, M., Reijers, H.A.: The multi-perspective process explorer. BPM (Demos) **1418**, 130–134 (2015)
22. Mansar, S., Reijers, H.A.: Best practices in business process redesign: use and impact. Bus. Process. Manag. J. **13**(2), 193–213 (2007)
23. Martin, P.Y., Turner, B.A.: Grounded theory and organizational research. J. Appl. Behav. Sci. **22**(2), 141–157 (1986)
24. Myers, M.D., Newman, M.: The qualitative interview in is research: examining the craft. Inf. Organ. **17**(1), 2–26 (2007)
25. Newman, I., Lim, J., Pineda, F.: Content validity using a mixed methods approach: its application and development through the use of a table of specifications methodology. J. Mix. Methods Res. **7**(3), 243–260 (2013)
26. Partington, A., Wynn, M.T., Suriadi, S., Ouyang, C., Karnon, J.: Process mining for clinical processes: a comparative analysis of four Australian hospitals. ACM Trans. Manag. Inf. Syst. **5**(4), 19 (2015)
27. Premkumar, G., Ramamurthy, K., Saunders, C.S.: Information processing view of organizations: an exploratory examination of fit in the context of interorganizational relationships. JMIS **22**(1), 257–294 (2005)
28. Reijers, H., Mansar, S.: Best practices in process redesign: an overview and qualitative evaluation of successful redesign heuristics. Omega **33**(4), 283–306 (2005)
29. del Rio-Ortega, A., Resinas, M., Cabanillas, C., Ruiz-Cortes, A.: On the definition and design-time analysis of process performance indicators. Inf. Syst. **38**(4), 470–490 (2012)

30. Urquhart, C., Fernández, W.: Using grounded theory method in information systems: the researcher as blank slate and other myths. In: Willcocks, L.P., Sauer, C., Lacity, M.C. (eds.) Enacting Research Methods in Information Systems: Volume 1, pp. 129–156. Springer, Cham (2016). https://doi.org/10.1007/978-3-319-29266-3_7
31. Venkatesh, V., Brown, S.A., Sullivan, Y.W.: Guidelines for conducting mixed-methods research: an extension and illustration. J. Assoc. Inf. Syst. 17(7), 435–494 (2016)
32. Wynn, M.T., et al.: ProcessProfiler3D: a visualisation framework for log-based process performance comparison. Decis. Support Syst. 100, 93–108 (2017)

Management

Effect of Attribute Alignment on Action Sequence Variability: Evidence from Electronic Medical Records

Inkyu Kim[1]([⊠]), Brian T. Pentland[1], Julie Ryan Wolf[2], Yunna Xie[3], Kenneth Frank[4], and Alice P. Pentland[2]

[1] College of Business, Michigan State University, East Lansing, MI, USA
{kiminky1, pentlan2}@msu.edu
[2] Dermatology, University of Rochester, Rochester, NY, USA
[3] Public Health Sciences, University of Rochester, Rochester, NY, USA
[4] College of Education, Michigan State University, East Lansing, MI, USA

Abstract. Business process mining algorithms discover processes from event logs that record sequences of events or actions. Typical event logs may or may not contain information about the *attributes* of the actions, such as the particular workstations used to carry out an action or the identity of the person performing the action. In this paper, we test the effect of action attributes on action sequence using data from electronic medical records at five dermatology clinics. We demonstrate that action sequence is influenced by attributes such as actors (who does what) and workstations (what is done where) that are not typically considered relevant to process flow control. We introduce a new metric – *attribute alignment* – that summarizes the extent to which actions are carried out with the same attributes throughout a process instance. If each action is always performed with the same attributes, attribute alignment is 100%. We discuss the implications and limitations of this finding for research and practice.

Keywords: Attribute alignment · Action sequence · Electronic medical records

1 Introduction

In process mining [1–3], processes are discovered from event logs that contain a stream of time-stamped actions or events [1, 4]. In a standard XES event log [5], actions may be associated with a set of *attributes*, such as an actor, machine, or location, but the discovered process is represented in terms of the actions. This makes sense because a

This material is based upon work supported by the National Science Foundation under Grant No. SES-1734237. Any opinions, findings, and conclusions or recommendations expressed in this material are those of the author(s) and do not necessarily reflect the views of the National Science Foundation. This research was also supported in part by University of Rochester CTSA (UL1 TR002001) from the National Center for Advancing Translational Sciences (NCATS) of the National Institutes of Health (NIH). The content is solely the responsibility of the authors and does not necessarily represent the official views of the National Institutes of Health. We are grateful for comments from Jan Recker and Jan Mendling on an early version of the analysis.

© Springer Nature Switzerland AG 2019
T. Hildebrandt et al. (Eds.): BPM 2019, LNBIP 360, pp. 183–194, 2019.
https://doi.org/10.1007/978-3-030-26643-1_11

process is a coherent, chronological sequence of interdependent events or actions [1, 4, 6].

In this paper, we investigate the effects of attributes such as actor and location on observed sequences of action in dermatology clinics. Using data from five dermatology clinics at the University of Rochester Medical Center (URMC), we examine the effects of *attribute alignment* on the sequences of action in clinical record keeping. Attribute alignment is a new construct that indicates the extent to which particular actions are consistently performed by the same actor in the same location. Here, we use attribute alignment as an indicator of the extent to which organizational roles (who does what) are consistently defined and carried out. When alignment is low, anyone can do anything, anywhere.

Contrary to our expectations, we find that attribute alignment has a stronger effect on action sequence than the clinic organization or the service performed in the particular visit. We use this finding as a basis for theorizing about how attributes can influence organizational processes.

While the contribution of the findings we report here is primarily theoretical, this research has an important practical motivation: the increasing cost and complexity of healthcare. This paper is part of a three-year research project that seeks to identify the antecedents of complexity in healthcare routines (NSF SES-1734237). Among other things, the research examines managerial factors such as clinical roles and organization. Preliminary results indicate that differences in how clinical roles are defined has a significant effect on process complexity [23]. In particular, clinics with "team documenters" (nursing staff who are responsible for maintaining patient records) have lower complexity than clinics where that responsibility is shared). Here, we dive into the underlying mechanisms that may help explain this phenomenon.

2   Theory

Research on organizational processes and routines naturally tends to focus on actions and patterns of action, because processes are described and defined in this way [6]. The focus is on the actions. In this section, we consider how the business process management incorporates (or excludes) context and attributes.

2.1   Layers of Context in Business Process Management (BPM)

In the BPM literature, there is interest in the role of context and in context-aware processes [7–12]. To help sort out the effect of context, Rosemann, Recker, and Flender [9] offer the "onion model", which consists of four layers of context: immediate, internal, external, and environmental.

Layers of the contextual onion tend to have different time scales relative to the cycle time of the focal process. Inner layers vary more quickly, while outer layers vary more slowly. Action attributes from the XES event log, such as the specific actor or workstation associated with an action, can be thought of as part of the immediate context of process execution [9]. Because action attributes (such as actor or workstation) can potentially vary with each action, *immediate* context can vary the fastest.

The internal context of a process might include the sequential relationship of actions within a sequence, as in [10]. Because internal context is relative to other parts of a process, it can also vary during each process instance (i.e., during each patient visit). In contrast, external context might vary by weekday/weekend, and environmental context might vary by time of year. External and environmental context tend to remain constant during any single process instance.

There is increasing interest in the analysis and design of context-aware processes [7, 10]. The effects of external and environmental context on process execution can be conceptualized and modeled using flow-control variables. For example, the execution of a car rental process may be based on location (airport vs. city pickup), season (winter vs. summer), and other contextual factors. Generally speaking, however, immediate contextual attributes such as who is performing an action, or which workstation is being used, are not considered relevant to flow control. These contextual details may or may not be present in the event log, and may not be included in the process model. The result is an action-only model that conforms the conceptual definition of a process [1, 4, 6], but leaves out the immediate context of process execution.

2.2   Task Design: Task Qua Task

Action-only models are entirely consistent with research on task design, where tasks are defined separately from the actor performing the task [13–15]. The phrase *task qua task* refers to the abstract idea of the task, separate from the execution of the task [15]. Research in this tradition advocates separating task from context as an explicit methodological principle [15] to avoid conflating properties of the task with properties of the people performing the task.

2.3   Action in Context

In contrast to the action-only approach, there are well established research traditions that emphasize the importance of context in the definition and interpretation of actions. For example, the pragmatic force of speech acts [16, 17] always depends on context. Expressions such as "here" and "now" mean something different depending on where and when they are uttered. More recently, theories of situated action [18, 19] make a strong argument for the importance of understanding the immediate context of an action. This leads us to expect that action attributes should not be overlooked when analyzing patterns of action.

In summary, there are theoretical reasons to expect that immediate context matters, but established theory in task design and current process mining methods generally do not consider action attributes when describing a process or a task.

3   Investigating the Effect of Attributes on Action Sequence

Event log data from an electronic medical record system provides an opportunity to explore the effects of immediate context empirically. We are particularly interested in understanding how action attributes, such as actors and locations, might influence the sequence of action in clinical work. To address this question, we need simple indicators that can be computed and compared using event logs with millions of observations. In the analysis that follows, we operationalize each of these constructs at the level of the process instance (one patient visit to one of the dermatology clinics).

Operationalizing Action Sequence. The most basic unit of sequential information is the 2-gram. In our usage, 2-grams represent pairs of sequentially adjacent actions in an event log. The number of unique 2-grams in a corpus of sequential data, such as a patient visit, is an indicator of how much sequential variety is present. If there are more unique 2-grams, there is more variety. Note that in principle, the number of unique 2-grams is independent of sequence length, because a sequence could consist of a single 2-gram repeated many times: a, a, a, a, a, a, a. In practice, we expect that longer sequences will have a larger number of unique 2-grams because greater length provides greater opportunity for variation. In the analysis that follows, the dependent variable is the number of unique 2-grams observed in an action sequence.

Operationalizing Attributes. To operationalize the role of attributes, we introduce a new construct that we call attribute alignment. It expresses the extent to which attributes add information to the description of an action. If the same actor always does the same action at the same workstation, then attribute alignment = 1. In this idealized case, knowing the action (or the actor, or the location) would give perfect information about the other attributes. The other attributes would be irrelevant.

In contrast, if multiple actors can perform a given action in multiple locations, then attribute alignment is low. Knowing the action does not determine the actor or the location (or vice versa). Attribute alignment provides a single number that encompasses the diversity of attributes associated with each action. The more diversity of attributes observed, the lower the alignment. Attribute alignment can be computed as follows:

$$\text{Attribute Alignment} = \frac{\text{Number of unique actions}}{\text{Number of unique action} - \text{attribute n} - \text{tuples}} \quad (1)$$

In the data we analyze here, we computed attribute alignment using action, role and workstation. To gain intuition for how this index works, consider a hypothetical example with three actions, three roles and three workstations. In the perfect alignment case, each unique action is performed by a single role at a specific workstation. The attribute alignment would be the maximum (1.0). In the low alignment case, there might be 27 distinct combinations of action-role-workstation ($3 \times 3 \times 3$). In that case, the alignment would be $3/27 = 0.11$.

Using these constructs, we can state a simple null hypothesis that we test in the following sections. We state this hypothesis in terms of correlation, rather than causality, because the constructs are operationalized within each process instance, so we cannot establish a definitive causal direction.

H_0: Attribute alignment is not correlated with action sequence.

This hypothesis reflects the idea that action sequences should be independent of the immediate context of task performance [7] and the influence of non-control flow variables. Stated in more theoretical terms, it reflects the idea that the *task qua task* exists independently of the actors performing the task and other attributes in the immediate context.

4 Methodology

4.1 Source of Data

Data was extracted from the EPIC Electronic Medical Record (EMR) audit trail at the University of Rochester Medical Center (URMC). This data traces actions in the EMR record keeping process. The data included two full years of patient visits from five dermatology clinics (over 7.7 million time-stamped records that provide a trace of actions for 57,836 patient visits, from January 2016 through December 2017). Descriptive features of the data are shown in Table 2 (below).

4.2 Example of Data

Table 1 provides an example of the data from the first five minutes of one visit. In addition to the time-stamped action, it contains a number of contextual factors: the role (e.g., admin tech), the workstation, the diagnosis and clinic ID. The role and workstation can be interpreted as immediate context. Note that some actions (e.g., MR_REPORTS) can be performed by any role at any workstation, so the attribute alignment for this visit will be less than perfect. The rows in Table 1 are shaded to show how the immediate contextual factors change throughout a visit, even at the level of individual actions. In contrast, Diagnosis and Clinic ID could be interpreted as external contextual factors. They remain the same throughout the visit.

4.3 Measurement of Variables

Unique 2-grams. The dependent variable in our analysis is the number of *unique* pairs of sequentially adjacent actions in a patient visit. To count unique 2-grams, we treated each patient visit as a sequence of actions. We identified 2-grams in each visit using the R package *n-gram*. We then counted the number of 2-grams that are unique. In any given visit, some 2-grams appear more than once, so the number of *unique* 2-grams is always lower than the length of the sequence.

Attribute Alignment. Attribute alignment is the number of unique actions in a visit divided by the number of unique action-role-workstation 3-tuples in the same visit. Each quantity is counted for each visit, so each visit has a value for attribute alignment.

Control Variables. We also control for a number of other factors that we expect to influence action sequences in the clinical record-keeping process.

- *Length of sequence.* Visits with more actions are likely to have more unique pairs of action, so we control for visit sequence length.
- *Clinic ID.* We know that each clinic has somewhat different procedures, so we include a dummy variable for Clinic.

Table 1. Example data

Time	Action	Role	Work-Station	Diagnosis	Clinic ID
2/2/15 8:53	CHECKIN TIME	Admin Tech	W1	Neoplasm	A
2/2/15 8:53	MR_SNAPSHOT	Admin Tech	W1	Neoplasm	A
2/2/15 8:53	MR_REPORTS	Admin Tech	W1	Neoplasm	A
2/2/15 8:53	MR_SNAPSHOT	Admin Tech	W1	Neoplasm	A
2/2/15 8:53	MR_REPORTS	Admin Tech	W1	Neoplasm	A
2/2/15 8:55	MR_SNAPSHOT	Admin Tech	W1	Neoplasm	A
2/2/15 8:55	MR_REPORTS	Admin Tech	W1	Neoplasm	A
2/2/15 8:56	MR_SNAPSHOT	Admin Tech	W1	Neoplasm	A
2/2/15 8:56	MR_REPORTS	Admin Tech	W1	Neoplasm	A
2/2/15 8:56	AC_VISIT_NAVIGATOR	Lic.Nurse	W3	Neoplasm	A
2/2/15 8:56	MR_HISTORIES	Lic.Nurse	W3	Neoplasm	A
2/2/15 8:56	MR_ENC_ENCOUNTER	Lic.Nurse	W3	Neoplasm	A
2/2/15 8:56	MR_VN_VITALS	Lic.Nurse	W3	Neoplasm	A
2/2/15 8:56	MR_REPORTS	Lic.Nurse	W3	Neoplasm	A
2/2/15 8:56	FLOWSHEET	Lic.Nurse	W3	Neoplasm	A
2/2/15 8:56	MR_VN_CHIEF_COMPLAINT	Lic.Nurse	W3	Neoplasm	A
2/2/15 8:56	MR_REPORTS	Lic.Nurse	W3	Neoplasm	A
2/2/15 8:56	MR_SNAPSHOT	Lic.Nurse	W3	Neoplasm	A
2/2/15 8:56	MR_REPORTS	Lic.Nurse	W3	Neoplasm	A
2/2/15 8:57	MR_REPORTS	Admin Tech	W1	Neoplasm	A
2/2/15 8:57	MR_SNAPSHOT	Admin Tech	W1	Neoplasm	A
2/2/15 8:58	MR_REPORTS	Lic.Nurse	W2	Neoplasm	A
2/2/15 8:58	AC_VISIT_NAVIGATOR	Lic.Nurse	W2	Neoplasm	A
2/2/15 8:58	MR_ENC_ENCOUNTER	Lic.Nurse	W2	Neoplasm	A
2/2/15 8:58	MR_HISTORIES	Lic.Nurse	W2	Neoplasm	A
2/2/15 8:58	MR_REPORTS	Lic.Nurse	W2	Neoplasm	A
2/2/15 8:58	MR_VN_VITALS	Lic.Nurse	W2	Neoplasm	A
2/2/15 8:58	FLOWSHEET	Lic.Nurse	W2	Neoplasm	A
2/2/15 8:58	MR_REPORTS	Physician	W4	Neoplasm	A
2/2/15 8:58	MR_VN_VITALS	Lic.Nurse	W2	Neoplasm	A
2/2/15 8:58	MR_HISTORIES	Lic.Nurse	W2	Neoplasm	A
2/2/15 8:58	MR_HISTORIES	Lic.Nurse	W2	Neoplasm	A
...	...	...	...	...	...

- *Level of service.* The Level of Service is a measure of the complexity of the service provided to the patient. It is used for billing and insurance, so it is based on auditable, objective factors about the patient visit.
- *Number of procedures.* This is the number of medical procedures performed during the visit. A typical procedure in a dermatology clinic would be freezing a wart.

Table 2. Descriptive statistics

Variables	N	Mean	St. Dev.
Unique 2-grams	57,784	89.27	23.75
Length of sequence	57,784	133.75	45.00
Level of service	55,294	3.05	0.38
Number of actions	57,784	36.70	7.64
Number of roles	57,784	3.33	1.01
Number of workstations	57,784	4.30	1.45
Number procedures	57,784	0.67	1.29
Attribute alignment	57,784	0.65	0.11

- *Number of actions.* Number of distinct actions during the visit. In the data set as a whole, there are 300 possible actions.
- *Number of roles.* Number of distinct actions during the visit. In the data set as a whole, there are 8 roles.
- *Number of workstations.* Number of distinct workstations during the visit. Across all four clinics, there were 118 workstations.

5 Findings

Table 3 shows the results of four regression models. In each model, the number of unique 2-grams is the dependent variable. To correct for heteroskedasticity, we ran our analysis with robust standard error. Due to the large sample size, all of the effects are statistically significant. To facilitate interpretation of the results, we report standardized coefficients and introduce the variables in groups. Note that our findings do not depend on whether the incremental R^2 from one model to the next is significant. Rather, we are interested in the relative size of the effects in the full model, as indicated by the magnitude of the standardized coefficients in model (4). In particular, we are concerned with the effect of attribute alignment, after controlling for everything else.

Model (1) shows the effect of attribute alignment, controlling for the length of the sequence. As expected, the length of the visit sequence is the dominant effect on the number of distinct 2-grams. The length of the visit alone accounts for 88% of the variance (adjusted $R^2 = 0.887$). Longer visits have many more unique 2-grams than short visits. Together with attribute alignment, the length of the visit accounts for nearly 90% of the variance in the number of unique 2-grams.

Model (2) controls for the effect of clinic organization and work practices by adding dummy variables for each clinic. We know that some clinics had dedicated staff that enter EMR data. In other clinics, residents and physicians do more of the recordkeeping work. As expected, clinic organization has a significant impact on action sequence.

Model (3) controls for the effect of the medical work as indicated by the level of service and the number of procedures. Some clinical visits are simple follow-ups to check if a condition is improving. Other clinical visits involve multiple procedures and tests.

Table 3. Number of unique 2-grams per visit (standardized coefficients)

Variables	(1)	(2)	(3)	(4)
Intercept	0.000 *** (0.376)	0.000 *** (0.415)	0.000 *** (0.480)	0.000 *** (0.403)
Attribute alignment	−0.101 *** (0.418)	−0.098 *** (0.444)	−0.098 *** (0.458)	−0.171 *** (0.388)
Length of sequence	0.873 *** (0.001)	0.873 *** (0.001)	0.870 *** (0.001)	0.690 *** (0.001)
Clinic 1		0.011 *** (0.218)	0.014 *** (0.230)	0.006 *** (0.180)
Clinic 2		0.009 *** (0.116)	0.016 *** (0.121)	0.007 *** (0.096)
Clinic 3		0.010 *** (0.099)	0.019 *** (0.103)	0.021 *** (0.082)
Clinic 4		0.012 *** (0.103)	0.020 *** (0.107)	0.012 *** (0.089)
Level of service			−0.006 *** (0.089)	−0.008 *** (0.069)
# of procedures			0.008 *** (0.026)	0.001 *** (0.020)
# of actions				0.282 *** (0.005)
# of roles				−0.007 *** (0.032)
# of workstations				−0.070 *** (0.026)
R^2	0.892	0.892	0.892	0.934
Adjusted R^2	0.892	0.892	0.892	0.934

$* \ p < 0.05, \ ** \ p < 0.01, \ *** \ p < 0.001$

These effects are statistically significant, but the magnitude is quite small. This appears to be because the recordkeeping work is not directly proportional to the actual clinical work.

Model (4) controls for the number of actions, roles and workstations observed in each visit, in addition to all of the prior effects. We add these controls to check if the mere number of actions, roles or workstations can account for the effect of attribute alignment. As expected, visits with more actions have more unique pairs of actions. Interestingly, visits with more roles and workstations have slightly fewer unique pairs of actions.

Across all of these models, we find a common result: as the attribute alignment goes down, the number of unique 2-grams increases. In other words, in visits where the attribute alignment is lower, the variation in sequence is higher. We have checked this result in many different ways (adding and removing other control variables, and aggregating the data in various ways). The result is robust. This leads us to reject the null hypothesis that action attributes do not affect action sequences.

6   Discussion

Who does what, and where they do it, has a substantial effect on the sequence of actions in these dermatology clinics. When clinical record keeping is carried out with greater alignment, it has less sequence variety. When alignment is lower, there is more sequence variety. After controlling for sequence length, attribute alignment is the single largest influence on sequence variety. Its effect is larger than clinic organization or the complexity of the work.

This finding is interesting because the dependent variable – the number of unique pairs of actions – is based *only on the sequences of actions*, regardless of who performs them or where they are performed. This leads us to suspect that the idealized *task qua task* [20], independent of who and where it is performed, does not exist in these dermatology clinics.

The influence of workstation is interesting because when a user logs in to the EPIC system, the screen is configured for that user. From the point of view of the users, every workstation is identical. Thus, we interpret the workstation as indicating the location of the work (e.g., in the examination room, at the nurses' station in the hall, in the front office, etc.). However, although personalization of user interface makes the workstation digitally identical, the effect of workstation may not be surprising because the physical environment of hospital could determine its influence. A busy hallway is different than a private office. Of course, these contextual differences are not generally conceptualized as relevant to process execution, but our study suggests that they can be.

The implication is that taking a particular action (e.g., *check_meds*) takes on a different meaning depending on who performs it and where it is performed. The office staff can *check_meds* at the workstation in the front office. This might be in response to a patient question (e.g., can I refill this prescription?). This might occur as the patient is checking in or checking out. Alternatively, a nurse, resident or doctor might *check_-meds* in the examination room, or outside the examination room, in order to confirm the dosage, look for conflicts, or write a new prescription. The point is obvious once we point it out: When the physician checks the patient's medication, it has a different significance than when the office staff does so. It looks like the same action in the event log, but it is not, because the immediate context is different.

6.1   Why Do Action Attributes Influence Action Sequence?

Intuitively, we did not expect action attributes and immediate context to influence action sequence. We expected that the structure of the work would determine the sequence of actions in the event log. In retrospect, we realize the error in our thinking: the "actions" in the event log are not fully defined by the action code. The logic here is simple. If *check_meds(physican, exam room)* is different than *check_meds(staff, front office)*, then the lexicon of actions in the real work is larger than the lexicon of actions in the event log. If the lexicon is larger, the number of unique 2-grams could be larger, as well. By omitting aspects of the immediate context, we are masking valid signals about the nature of the work.

6.2   Including Attributes in the Definition of Actions

Abstracting away contextual details can produce cleaner, more general models, but our findings suggest that this may be a mistake in some cases. Rather than suppressing the immediate context, perhaps we should find ways to include it in our models?

Towards that end, Pentland et al. introduced *ThreadNet*, a simple tool for visualizing and analyzing routines and processes [21, 22]. *ThreadNet* is an *R* package that can be downloaded from GitHub (https://github.com/ThreadNet/ThreadNet). *ThreadNet* provides a convenient interface for defining nodes in the graph in terms of any number of attributes. Thus, action attributes become part of the model. When attribute alignment is low, this does tend to result in a proliferation of nodes. However, our results here indicate that the additional complexity may provide empirical insights that an action-only perspective would miss. For example, a process model that includes roles and workstations may help us understand the effect of clinic organization on outcomes such as process complexity and patient satisfaction [23].

6.3   Limitations

This is, in effect, a case study of EMR record-keeping in five dermatology clinics, all operating in the same hospital network. Thus, we should not over-generalize from these findings. In other contexts, the sequential structure of the *task qua task* may be impervious to who is doing the work, where it is performed, or other action attributes.

Using the number of unique 2-grams as the basis for comparison is simple, but two different process instances might have the same number of unique 2-grams. As a result, we believe this metric tends to understate the phenomenon it is intended to measure. It might be more informative to use optimal matching or some other methodology to measure sequential variety [10, 22].

Finally, our contribution at this stage is primarily theoretical. We have shown that attribute alignment can have an unexpected effect on process execution, but we have not yet connected this theoretical insight to practical outcomes, such as cost, quality, or satisfaction.

7   Conclusion

This research demonstrates that, at least in these dermatology clinics, action attributes that are not normally considered relevant to process execution can influence observed sequences of action. To demonstrate this effect, we have introduced a novel measure that we call attribute alignment. We suggest that future research should capture more detailed, event-level contextual information so that the managerial implications of these effects can be investigated more broadly.

References

1. Van Der Aalst, W.: Process Mining: Discovery, Conformance and Enhancement of Business Processes. Springer, Heidelberg (2011). https://doi.org/10.1007/978-3-642-19345-3
2. Breuker, D., Matzner, M., Delfmann, P., Becker, J.: Comprehensible predictive models for business processes. MIS Q. **40**, 1009–1034 (2016)
3. Dumas, M., La Rosa, M., Mendling, J., Reijers, H.A.: Fundamentals of Business Process Management. Springer, Heidelberg (2013). https://doi.org/10.1007/978-3-642-33143-5
4. Van der Aalst, W., Weijters, T., Maruster, L.: Workflow mining: discovering process models from event logs. IEEE Trans. Knowl. Data Eng. **16**, 1128–1142 (2004)
5. Acampora, G., Vitiello, A., Di Stefano, B., Van Der Aalst, W., Gunther, C., Verbeek, E.: IEEE 1849: the XES standard: the second IEEE standard sponsored by IEEE computational intelligence society [society briefs]. IEEE Comput. Intell. Mag. **12**, 4–8 (2017)
6. van der Aalst, W., et al.: Process mining manifesto. In: Daniel, F., Barkaoui, K., Dustdar, S. (eds.) BPM 2011. LNBIP, vol. 99, pp. 169–194. Springer, Heidelberg (2012). https://doi.org/10.1007/978-3-642-28108-2_19
7. Rosemann, M., Recker, J.C.: Context-aware process design: exploring the extrinsic drivers for process flexibility. In: The 18th International Conference on Advanced Information Systems Engineering. Proceedings of Workshops and Doctoral Consortium, pp. 149–158. Namur University Press (2006)
8. Günther, C.W., Rinderle-Ma, S., Reichert, M., Van Der Aalst, W.M., Recker, J.: Using process mining to learn from process changes in evolutionary systems. Int. J. Bus. Process Integr. Manag. Spec. Issue Bus. Process Flex. **3**, 61–78 (2008)
9. Rosemann, M., Recker, J.C., Flender, C.: Contextualisation of business processes. Int. J. Bus. Process Integr. Manag. **3**, 47–60 (2008)
10. Bose, R.J.C., Van der Aalst, W.M.: Context aware trace clustering: towards improving process mining results. In: Proceedings of the Ninth SIAM International Conference on Data Mining, pp. 401–412. SIAM (2009)
11. Van Der Aalst, W.M., Dustdar, S.: Process mining put into context. IEEE Internet Comput. **16**, 82–86 (2012)
12. Anastassiu, M., Santoro, F.M., Recker, J., Rosemann, M.: The quest for organizational flexibility: driving changes in business processes through the identification of relevant context. Bus. Process Manag. J. **22**, 763–790 (2016)
13. Wood, R.E.: Task complexity: definition of the construct. Organ. Behav. Hum. Decis. Process. **37**, 60–82 (1986)
14. Campbell, D.J.: Task complexity: a review and analysis. Acad. Manag. Rev. **13**, 40–52 (1988)
15. Hackman, J.R.: Toward understanding the role of tasks in behavioral research. Acta Physiol. **31**, 97–128 (1969)
16. Austin, J.L.: How to Do Things with Words. Oxford University Press, Oxford (1975)
17. Searle, J.R., Searle, J.R.: Speech Acts: An Essay in the Philosophy of Language. Cambridge University Press, Cambridge (1969)
18. Suchman, L.A.: Plans and Situated Actions: The Problem of Human-Machine Communication. Cambridge University Press, Cambridge (1987)
19. Feldman, M., Orlikowski, W.: Theorizing practice and practicing theory. Organ. Sci. **22**, 1–14 (2011)
20. Hærem, T., Pentland, B.T., Miller, K.D.: Task complexity: extending a core concept. Acad. Manag. Rev. **40**, 446–460 (2015)

21. Pentland, B.T., Recker, J., Wyner, G.: Rediscovering handoffs. Acad. Manag. Discov. **3**, 284–301 (2017)
22. Pentland, B., Recker, J., Kim, I.: Capturing reality in flight? Empirical tools for strong process theory. In: Proceedings of Thirty Eighth International Conference on Information Systems, pp. 1–12 (2017)
23. Ryan, J.L., Xie, Y., Kim, I., Frank, K., Pentland, A.P., Pentland, B.T.: Team documentation influences clinic complexity and patient satisfaction. J. Invest. Dermatol. **139**(5), S106 (2019)

From Openness to Change to Patients' Satisfaction: A Business Process Management Approach

Yevgen Bogodistov, Jürgen Moormann[(⊠)], and Rainer Sibbel

Frankfurt School of Finance and Management, ProcessLab, Adickesallee 32-34,
60322 Frankfurt, Germany
{y.bogodistov, j.moormann, r.sibbel}@fs.de

Abstract. On the example of the current health-care reform in Ukraine, the impact of process changes on the level of individual providers of care is analyzed. The empirical study focuses on the question of how much effective Business Process Management (BPM) and openness to change can help health-care providers like family doctors to keep their performance during a structural transformation. The theoretical framework of the study is based on an innovative human-centric approach to BPM. The results show that openness to change reduces the risk of burnout and reduces cooperation issues with other doctors and hospital departments. At the same time, burnout and cooperation issues increase the number of errors and the time span of diagnosing patients significantly.

Keywords: Business Process Management · Process change ·
Job demands–resources model · Health-care management

1 Introduction

Nowadays, organizations are confronted with permanent change such as globalization, political realignments, and the rapid advance of information technology. In the health-care sector, due to demographic trends, medical and technological progress, as well as regulatory and economic pressure, change is of crucial importance for all groups of stakeholders. This includes in particular those who are employed in this sector and those who are the receivers of health-care services – the patients [1–3]. From the viewpoint of Business Process Management (BPM) a reform can have many faces: it can appear as a number of relatively simple process improvements or can be concerned with structural process innovation, leading to a drastic shift in roles, responsibilities, and the way processes are conducted. If processes have to be changed completely, one can assume a higher degree of resource investment on the side of the health-care provider and from its personnel. The health-care provider has to develop new processes, adapt its organizational structure, provide training and seminars, and support employees in their acceptance of change (e.g., [4]). The personnel also have to contribute, as they have to invest their psychological and physical resources [5].

© Springer Nature Switzerland AG 2019
T. Hildebrandt et al. (Eds.): BPM 2019, LNBIP 360, pp. 195–210, 2019.
https://doi.org/10.1007/978-3-030-26643-1_12

One of these change domains can be found when a health-care reform is performed, for instance a recent comprehensive reform in Ukraine. According to a World Health Organization report [6], the reform of the Ukrainian National Health Service has to be accompanied by a decentralized structure of service delivery, a new way of financing, and so on. In such an event, one can observe a series of substantial changes or even process innovations [7]. First, by separation of the providers for primary and secondary health care, the areas of responsibility shift – family doctors or general practitioners receive more responsibility, since they become the first line of inquiry. In addition, their interaction extends beyond specific episodes of illness or disease and thus is based on a long-term care approach [8, 9]. Second, usually primary and secondary health care require different forms of financing, which cause different physician–patient relationships (e.g., [10]). Third, family doctors need to have special education allowing them to diagnose even complex illnesses at the first line of patient–doctor interaction. Indeed, several studies show that there are ways to diagnose eating disorders [11], risk of suicide [12], or depression [13], which, if introduced to primary care, could unburden the health-care system.

In this work, we focus on publicly owned and operated health-care providers in Ukraine. The decisions concerning organizational transformation are mostly made on the governmental level (e.g., Ministry of Health Care). For instance, hospitals might not be able to freely decide on the usage of resources. Moreover, they have a set of requirements, legal regulations, and standards conferred upon them. Indeed, the government becomes the entity responsible for, first, the evaluation of opportunities; second, the direction of transformation; and, third, the control of implementation.

Reforming the health-care system through the introduction of primary health care reduces costs and improves health equity [14, 15]. However, the introduction of primary care is a complex endeavor. In order to cope with such a transition, health-care providers need to learn how to manage organizational processes in an efficient and effective manner. Several established routines, processes, and procedures will remain unchanged (e.g., blood tests, drugs prescribing, illness registration), whereas other processes (e.g., interaction with patients, health data analysis, internal communication) will change. Moreover, the administrative framework (e.g., financing), resources (e.g., requirements for personnel), and roles (e.g., who performs a blood test) will also shift.

Since investments can be made on the organizational (e.g., health-care provider) and the individual (e.g., employees) level, both parties are responsible for the efficiency and effectiveness of coping with the transition. But it is the health-care provider who has to deliver the framework for the transition. Therefore, we postulate the following research question: *How much can effective process management and openness to change help health-care providers perform better during a structural change?*

In order to answer this question, we apply the Job Demands–Resources (JD-R) model and test it in the domain of a structural reform – the separation into primary and secondary health-care systems. We root our theory in Business Process Management (BPM) and expand the most ignored part of it – the individuals – by means of a well-established theoretical framework from the field of work and organizational psychology. Further, we develop a new human-centric approach to BPM and support it with empirical data from a hard-to-access sample.

2 Theoretical Framework

2.1 Managing Process Change

All industries are confronted with massive changes. Therefore, companies and other organizations, heterogeneous in their resources, need to adapt their strategies and business processes in order to maintain or develop competitive advantage [16, 17]. It is against this background that the field of BPM gained wide attention in practice and theory (e.g., [18–20]). BPM does not only aim at improving some existing processes, but is a form of organizational change which is characterized by the strategic transformation of interrelated organizational subsystems, producing varied levels of impact [21].

During a structural reform, many processes need to be changed as well as new processes developed. This leads to huge resource investments. These investments happen at the organizational and the individual levels of an organization. The individual level is particularly important because the people (respectively the workers' council) must consent to process changes. If an organization provides sufficient information on change and allows for the participation of employees in decision making, the personnel show a higher degree of acceptance as well as job satisfaction, less work irritation, and less intention to quit [22]. Although BPM recognizes various components critical for BPM success, two of which are culture and people, the individual level is still largely neglected in the BPM literature [20, 23].

In their attempt to explain the issues of motivation, physiological (fatigue), and psychological (burnout) strain, psychologists proposed a framework incorporating different types of resources required from or offered to an employee [24, 25]. The model was called the Job Demands–Resources model, since it looked at the balance between the resources required and resources offered.

The JD-R model assumes that individuals have a set of psychological and physiological resources which they expend for various job demands. These resources stem from physical, psychological, social, or organizational aspects of the job and are associated with certain physiological or psychological costs [26]. Examples of organizational demands are unfavorable working environments, work pressure, and emotional demands. In order to meet these demands, an employee has to invest effort, for instance provide emotional labor in the form of deep (feeling emotions) or surface (showing emotions) acting [27]. An organization can help employees to meet these job demands by providing job resources, such as autonomy, more flexibility in work design, supervisor's support, and so on. Employees try to spend their psychological and physiological resources wisely and even tend to accumulate them [28]. If a process is poorly designed, it can exhaust the mental and physical resources of employees, resulting in disengagement, underperformance, or even burnout [26].

2.2 Structural Reform and Openness to Change

Process management in the highly centralized and regulated domain of health care is difficult to perform [29]. Nonetheless, the change is of crucial importance, since the lives of people are at stake.

A health-care reform is an example of drastic changes: the way of financing, the way of interaction with patients, as well as the way of internal communication and cooperation are objects of the transformation. At the same time, the extant system cannot disappear and be altered overnight. Thus, the patients have to be treated and the doctors have to keep working while their roles, tasks, and responsibilities are being changed. Such a transition produces enormous workloads and psychological stress. Indeed, organizational change becomes an antecedent to the affective states of employees [30]. For example, Dahl [31] shows that organizational change evokes stress, causing significant risks for employees' health. Wanberg and Banas [22] argue that coping with change for employees might be a difficult task, because employees might feel a loss of territory, and become uncertain concerning their future roles and the tasks they may face.

In practice, due to an external event (decision of the parliament or ministry) the existing path has to be abandoned, whereby path-dependency theory assumes the high costs of such changes. The theory assumes the investment of financial resources in the event of a radical change. We argue, however, that it is also about other types of investment an organization has to make.

Different types of process change are associated with different job demands. Figure 1 depicts four quadrants resulting from two dimensions – job demands (low/high) and job resources (low/high). We added a 'Degree of change' axis below the 'Job demands' axis. The higher the degree of change (e.g., process innovation), the higher the job demands. For instance, if employees experience substantial changes in their processes, they have to deal with a set of new job demands:

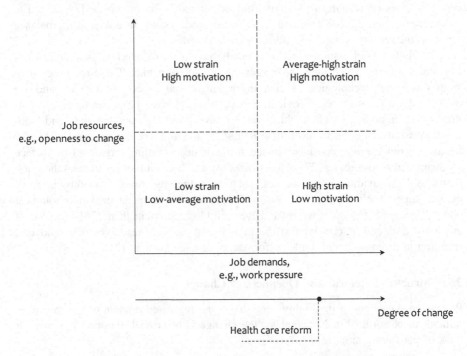

Fig. 1. Human-centric process change framework (developed based on [26]).

- new roles and responsibilities,
- additional time for training and mastering new skills, and
- accepting new technologies taking away their previous job contents.

A health-care reform would be associated with high job demands close to those of process innovation. Knowing this, process managers can identify the quadrants on the right of Fig. 1 they will need to deal with. If they do not offer sufficient job resources, they will end up in high strain (burnout) and low motivation of their staff members. However, if they offer more job resources (e.g., social support), they can reduce strain and even increase job motivation.

Indeed, social support is known as one of the buffers against job strain (e.g., [32]), which is apparently concerned with new processes. Without social support, organizations might face resistance to change from their employees, hampering the transition. Resistance can result in negative outcomes such as absenteeism, decreased satisfaction, or low productivity [33].

An organization's openness to change, we define, is a construct incorporating social, organizational, physiological, and psychological support of employees during the transition from the current state to the aspired state. Organizational openness to change relies on the policy an organization follows when (a) sharing information with its employees, (b) letting employees participate in decision making, and (c) helping employees cope with the change. Our definition is backed up by the work of Wanberg and Banas [22], who show that these three components are crucial for employees to accept change and its psychological (e.g., job satisfaction, work irritation) and behavioral (e.g., intent to quit) consequences.

In the case of a structural change such as a health-care reform, BPM scholars and practitioners would typically suggest applying operational excellence initiatives such as Total Quality Management, Theory of Constraints, Six Sigma, and Lean Management. By incorporating the JD-R model, we expand this approach. We argue that in order to deal with a structural transformation, an organization has to make additional investments of a rather psychological nature.

3   Research Model

As our theory suggests, by providing organizational support for change, employees receive additional job resources with which they can balance the high job demands during a structural reform. Indeed, Wanberg and Banas [22] show that sharing information on change, self-efficacy of employees in coping with change, and participation in decision making are positively associated with a higher level of change acceptance. Allen et al. [34] emphasize the role of trust and communication of information. Communication reduces uncertainty and therefore lowers stress predictors. Openness to change would allow for better communication, whereas organizational "saving on this resource" would undermine the acceptance of change by employees.

The explanation for the lack of acceptance of change by employees can be found in Hobfoll [28], who proposes the notion of resource conservation by employees; that is, employees tend to obtain, retain, and protect material, social, personal, or energetic resources they value. Consequently, if an organization is not open to change and thus does not provide additional job resources to its employees, these employees would start saving their psychological (e.g., less emotional involvement) or physiological (e.g., longer breaks or sleep at work) resources. For instance, emotional labor is one of the main predictors of burnout [27]; or, put differently, if an employee is required to show certain emotions, s/he might start saving this valuable resource and behave cynically. Schaufeli [5, p. 121] argues: "when job demands [...] are chronically high and are not compensated by job resources [...], employee's energy is progressively drained. This may finally result in a state of mental exhaustion ('burnout'), which, in its turn, may lead to negative outcomes for the individual (e.g., poor health) as well as for the organization (e.g., poor performance)."

The JD-R model, therefore, does not only explain the direct influence of a job resource such as openness to change on communication and cooperation, but also the indirect influence via burnout. This influence is expected to be inconsistent, since openness to change might *reduce* burnout. At the same time, burnout might *increase* the cooperation issues. Therefore, we hypothesize:[1]

> Hypothesis 1. Openness to Change has a negative influence on Cooperation Issues.
> Hypothesis 2. Openness to Change has an indirect inconsistent effect (via Burnout) on Cooperation Issues.

An interesting side effect of the JD-R model, combined with Hobfoll's [28] theory, is the notion of "gain" and "loss spirals." During the process change, employees need to invest their resources in order to prevent a loss of resources. Thus, those employees who have fewer resources (e.g., less support from their organization) cannot withstand turbulent environments and lose even more of their resources ("loss spiral"). On the contrary, those who have more resources will seek opportunities to invest them in order to increase their resource gains ("gain spiral"). Accordingly, those employees who have lost resources due to burnout or who have fewer resources due to a lack of organizational support will not be able to use cooperation in order to be more effective. Missing resources will result in an effect on the employees' performance: the number of errors in their work will increase which, hence, will lead to more time needed to accomplish their tasks.

In our research, we look at process management of health-care providers while they face a structural reform. In this regard, we focus on the core process of "diagnosing." This process is strongly impacted by the transformation. The separation into primary and secondary care means that the roles of doctors and their responsibilities in this process will shift: a family doctor receives more responsibilities and, thus, has higher job demands. If our theory is correct, cooperation issues (caused by lack of organizational support and burnout) will lead to an increased number of errors while diagnosing. A higher number of errors leads to longer pathways of communication (e.g., due to internal conflicts) and thus processing times. One of the reasons for these issues might also be burnout. The latter might lead to unwillingness of doctors to engage in

---

[1] Henceforth the names of the constructs are capitalized.

cooperation activities that might cause further errors in diagnosing due to a lower quality of information. Consequently, we hypothesize:

Hypothesis 3. Cooperation Issues increase Time Span of Diagnosing.
Hypothesis 4. The impact of Cooperation Issues on Time Span of Diagnosing is mediated by Errors in Diagnosing.
Hypothesis 5. Burnout increases Errors in Diagnosing.
Hypothesis 6. The impact of Burnout on Errors in Diagnosing is mediated by Cooperation Issues.

Our hypotheses are depicted in Fig. 2.

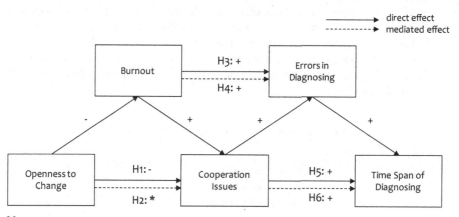

Note:
* stands for inconsistent mediation.

Fig. 2. Research model.

4 Methodology

4.1 Sample

In order to test our hypotheses, we collected data in two Ukrainian cities from the same region: Alpha (population approx. 650,000) and Bravo (population approx. 20,000). Data collection started in October and ended in November 2017. A total of 145 family doctors from the city of Alpha participated in our study, resulting in 71% of the population (the actual population in Alpha is 204 family doctors). Bravo is a small city with a population of 36 family doctors. We collected answers from 33 family doctors, resulting in 92% of the population. The fact that the family doctors participated in such big numbers indicates the high relevance of the topic and willingness to cooperate.

The overall sample comprises 179 family doctors, 141 of whom are female and 37 are male. The average age is 43.9 years (standard deviation [SD] = 13.83). The average working experience is 20.6 years (SD = 14.60), whereas the average working experience in the current hospital is 7.9 years (SD = 8.43).

4.2   Measures

Independent Variables. The independent variables in our study were Openness to Change (OTC) and Burnout. For burnout we used the Copenhagen Burnout Inventory (CBI, [35]). All items were assessed using a 7-point Likert-type scale, ranging from "(Almost) never/To a very small extent" to "(Almost) always/To a very high extent." We developed a construct of OTC, whereby we used a 7-point Likert scale, ranging from "Completely disagree" to "Completely agree." It is important to notice that our construct represents OTC as felt by the recipients.

Dependent and Mediating Variables. We developed a set of variables to capture the quality of processes in hospitals. All items were assessed using 7-point Likert scales, ranging from "Is decreasing" to "Is increasing." The construct Cooperation Issues was concerned with struggles family doctors encounter when cooperating with other doctors or hospital departments. It was measured with three items (Table 1). Errors in Diagnosing were also treated as a perceived measure. We used three reflective items to capture the construct. Time Span of Diagnosing was measured using four items. All items were reflective.

Table 1. Constructs with related items and factor loadings

Construct	Item#	Item	Loading
Burnout[a]		**Personal Burnout**	.920[b]
		Work-related Burnout	.960[b]
		Client-related Burnout	.847[b]
Openness to change	Item 1	My hospital is open to process change	.839
	Item 2	Personnel involved in the process change feel like an integral part of the hospital	.848
	Item 3	Members (doctors and administrative personnel) of the process change team(s) feel rewarded for innovations or ideas they have implemented	.765
	Item 4	The organizational structure of the hospital assists process change	.818
Cooperation issues	Item 1	Number of misunderstandings among departments	.867
	Item 2	Number of patients "ping-ponged" between departments	.828
	Item 3	Number of internal conflicts due to a shift in responsibilities	.848
Errors in diagnosing	Item 1	Number of patients with a wrong diagnosis	.744
	Item 2	Number of patients sent (back) by specialists because the diagnosis should be done on the primary level	.745
	Item 3	Number of patients with redundant treatment(s)	.808

(*continued*)

Table 1. (*continued*)

Construct	Item#	Item	Loading
Time span of diagnosing	Item 1	Time for diagnosing common illnesses (allergy, cold and flu, conjunctivitis, diarrhea, headache, stomach ache, etc.)	.710
	Item 2	Time for diagnosing other (less common) illnesses	.802
	Item 3	Time for diagnosing patients requiring basic medical tests	.794
	Item 4	Time for diagnosing patients requiring advanced medical tests	.832

Note: a We used items from Kristensen et al. [35], whereas the item #10 ("Do you have enough energy for family and friends during leisure time," reverse scored) was deleted due to its low factor loading of .047.

b Indicates loading of the first-order construct on the second-order construct of Burnout.

Control Variables. In our study we controlled for several effects. First, we controlled for the influence of age and gender on Burnout (e.g., [36, 37]). We also controlled for their influence on OTC. We assumed that age as well as experience might reduce OTC and Cooperation Issues. Additionally, we included measures on experience in the specific hospital, assuming that it might have an effect on the quality of cooperation, resulting in a lower Time Span for Diagnosing as well as in fewer Cooperation Issues. We also controlled for the influence of the number of employees and the number of family doctors in the hospital.

4.3 Reliability and Validity

Since we developed a set of items, we needed to perform a Confirmatory Factor Analysis (CFA). Moreover, we needed to ensure discriminant and convergent validity as well as the internal consistency of our constructs. We calculated Cronbach's α, Composite Reliability (CR), Average Variance Extracted (AVE), Maximum Shared Variance (MSV), and Maximum Reliability H (MaxR(H)), as well as average factor loadings. All indicators showed a high quality of the constructs [38] and can be found

Table 2. Reliability and validity statistics.

	α	CR	AVE	MSV	MaxR(H)	CBI	TSD	ED	CI	OTC
CBI	.954	.935	.828	.160	.952	**.910**[a]				
TSD	.792	.866	.618	.228	.964	.093	**.786**[a]			
ED	.649	.810	.587	.264	.969	.295	.478	**.766**[a]		
CI	.805	.885	.719	.264	.975	.400	.363	.514	**.848**[a]	
OTC	.832	.894	.678	.098	.979	−.244	−.095	−.111	−.313	**.823**[a]

Note: a Indicates average factor loadings, CBI – Burnout (Copenhagen Burnout Inventory), TSD – Time Span of Diagnosing, ED – Errors in Diagnosing, CI – Cooperation Issues, and OTC – Openness to Change.

in Table 2. Although Errors in Diagnosing showed a relatively low but acceptable Cronbach's α, according to Churchill's guideline [39], the additionally calculated CR did not indicate any problem and largely exceeded the common cut-off value.

We calculated a series of model fit indicators which confirmed a good model fit: $\chi^2/$ df is 1.751 (χ^2 = 754.645, df = 431), Comparative Fit Index (CFI) is .911, Root Mean Square of Approximation (RMSEA) is .065, Standardized Root Mean Square Residual (SRMR) is .0647 [40]. The variance explained by a common latent factor is 5.29%, confirming that common method bias can be neglected.

5 Results

In order to test our model, we applied the partial least squares algorithm (PLS) using SmartPLS™ software. As a second-generation structural equation modeling technique, it can estimate the loadings and weights of indicators on constructs (therefore estimating construct validity) and the causal relationships among constructs in a model [41]. In addition, PLS is most suitable for models with second-order constructs (in our case, Burnout) and relatively small samples [38].

We ran the full model with all mediating constructs and control variables. The R^2 for Cooperation Issues is .23, R^2 for Errors in Diagnosing is .28, and R^2 for Time Span of Diagnosing is .25, indicating an appropriate model fit [42]. The lowest R^2 of .10 is shown for Burnout, which we hold acceptable since the main predictors of Burnout are usually psychological factors such as emotional labor, surface action (e.g., [27, 43]), workload [e.g., 44], and stress [45].

All our control variables apart from age did not show significant relationships. Age, on the contrary, produced a weakly significant impact on Burnout ($\beta = -.192$,

Table 3. Results.

IV	MV	DV	β	t	p	Sobel test, p
OTC		CBI	−.244	2.116	.035	
OTC	–	CI	−.314	2.811	.006	
CBI		CI	.333	3.196	.001	
OTC	CBI	CI	−.226	2.048	.041	.069
CBI	–	ED	.279	2.723	.007	
CI		ED	.451	4.341	<.001	
CBI	CI	ED	.104	1.049	n.s.	.008
CI	–	TD	.371	3.871	<.001	
ED		TD	.393	3.342	<.001	
CI	ED	TD	.163	1.332	n.s.	.004

Note: IV – independent variable, MV – mediating variable, DV – dependent variable, CBI – Burnout, TSD – Time Span of Diagnosing, ED – Errors in Diagnosing, CI – Cooperation Issues, OTC – Openness to Change, n.s. – non-significant.

t = 1.765, p = .078) and a significant impact on Cooperation Issues (β = −.129, t = 1.982, p < .048). We refer to this finding in the discussion section of our work. In the further analysis we focus on the main effects, whereby all control variables remain in all further model calculations. The results can be found in Table 3.

In the full model, we found a statistically significant influence of Openness to Change on Burnout. As predicted in our hypothesis, OTC reduces Burnout of family doctors. At the same time, OTC reduces Cooperation Issues. With regard to Burnout, our findings reveal the following. First, Burnout has a strong and high impact on Cooperation Issues. As expected, Burnout of family doctors increases problems in their cooperation with other doctors and departments of the health-care provider. Second, in order to test the indirect effect of Openness to Change on Cooperation Issues (via Burnout) we performed a Sobel test [46], which supports our hypothesis 2, although on a p < .1 significance level. Since Openness to Change reduces Burnout, which increases Cooperation Issues, we assume an inconsistent mediation [47], and since the effect in the model with the mediator remains significant, the mediation is partial. Thus, our hypothesis 1 is partially supported.

Cooperation Issues increase the number of errors while diagnosing patients. Interestingly, in the full model Burnout did not produce a significant impact on Errors in Diagnosing. Since we assumed a mediation in this relationship, we ran the same model without the Cooperation Issue construct. The result became strongly significant. Therefore, hypothesis 5 is confirmed. A Sobel test supported the mediation: we found a statistically significant relationship. As can be seen from the previous tests, the impact of Burnout without Cooperation Issues as mediator increases by more than twice (β = .104 - > β = .279). Therefore, our hypothesis 6 is supported.

Then, we looked at the relationship between Cooperation Issues, Errors in Diagnosing, and Time Span of Diagnosing. Errors in Diagnosing show that more time is needed to diagnose patients. At the same time, Cooperation Issues does not show a significant effect on Time Span of Diagnosing. Since we assumed a mediation, we ran the same model without the construct Errors in Diagnosing. Indeed, Cooperation Issues increased Time Span of Diagnosing on a statistically significant level. Consequently, we ran the Sobel test to examine the significance of mediation effects. The test supported our expectations. As can be seen in previous tests, when introducing a mediator, the effect drops from β = .371 to β = .163; that is, more than twice. Consequently, our hypotheses 4 and 5 are supported.

The R^2 values reflect variance explained in the specific variable, but not the ability of parameter estimates to match the sample covariances [48]. To reassure us on the robustness of our model with regard to the quality of our latent variables, we used the covariance-based software IBM AMOS. We added all latent constructs and calculated the model fit indicators, which confirmed a very good model fit for the model: χ^2/df = 1.764 (χ^2 = 767.301, df = 435), CFI = .909, RMSEA = .065, SRMR = .0693 [40].

6 Discussion

At the beginning of this work, we asked ourselves the question: *How much can an effective process management and openness to change help health-care providers perform better during a structural change?* Our empirical data demonstrates that the influence is both complex and significant. We show that openness to change, which can also be referred to as the ability of an organization to involve employees in the process of transformation, has an effect on the process's outcome via a comprehensive set of relations. It appears that the openness to change of a hospital supports cooperation, which, in turn, helps reduce the number of errors or, put differently, fosters process quality. The reader has to bear in mind that our measurement of openness to change referred to the broader readiness of organizations to support change, and not to a specific process. This brings us to the conclusion that an open organizational culture, supporting and promoting change, has a positive effect on process quality [49]. One of the possible explanations we found is the level of strain, which is lower in open-to-change organizations. Since we have found only a partial mediation, a search for further explanatory variables is necessary. Furthermore, not only the quality of processes but also throughput times (here: time span of diagnosis) become shorter. Practically, we show that openness to change allows for expanding the "magic triangle": with constant costs (the number of family doctors and, thus, related employment costs are constant), the time and quality of processes improve.

While developing the paper we noticed that the introduction of the JD-R model into BPM has a much deeper impact on theory than we initially expected. The JD-R model is not only helping to incorporate the hitherto neglected psychological and physiological resources into process management, it also appears commensurable with BPM. Finding the balance between job demands and resources with regard to the specific individual resources might help create motivation, increase the success of projects, and reduce burnout and other types of psychological and physiological strain. The imbalanced job demands and resources might result in low quality or a time-delayed output – a result that does not appear from the regular BPM approach. The (negative) outcome might result in additional organizational costs or, as we show in our study, errors in diagnosing. Moreover, motivation, which is acknowledged by BPM [50, 51], should not be taken for granted but be understood as the outcome of an appropriate balance of job demands and resources. BPM, expanded by the results of our study, is hence of crucial importance.

For practitioners, our study implies the need to consider individual resources thoroughly in the context of BPM. Managers should acknowledge that processes need to be managed through the provision of specific job resources, such as moral support (psychological, social), flexible work schedules (physiological), or shorter decision paths due to less hierarchy (organizational job resources). Markos and Sridevi [52] analyze the two-way relationship between employee and employer and find that engaged employees are not only emotionally attached to their organization, but also willing to go the extra mile when needed. This "extra mile," according to our theory, reflects the motivation/strain balance and allows for inclusion of the latter into BPM. Process managers ignoring the role of individuals will increase strain that might cause burnout, disengagement, and other health problems [5]. Thus, the management of job resources, such as psychological, physiological, social, and organizational job demands caused by business processes, we argue, should be incorporated into all levels of process maturity models.

7 Limitations and Further Research

Several limitations of our study must be acknowledged. First, data on errors in diagnosing and time span of diagnosing are subjective assessments. Due to ethical reasons as well as legal limitations, we were not allowed to collect direct data on the time span of diagnosing or on the number of errors. Moreover, we do not think that doctors would report this data. We decided to go for a reflective measure of the doctors' perception of the overall diagnosing and interaction process. We were fascinated with the trust the doctors conferred upon our team when answering these questions.

Second, the reader has to bear in mind that we found only a partial inconsistent mediation between openness to change and cooperation issues (via burnout). We interpret this as follows: there might be other mediators that explain why openness to change reduces issues in cooperation between family doctors and other departments of the respective hospital. Additionally, we would like to stress the role of burnout in BPM. Our model produced a relatively low R^2 for burnout. However, this is not an issue, since we did not argue that openness to change is the main predictor of burnout. As the literature shows, the main predictors of burnout are emotional labor, surface action (e.g., [27, 43]), workload (e.g., [44]), red tape, interaction partners, and internal doubts [53]. An additional model fit test reassured us of a good model fit with regard to all the latent constructs used.

Third, generalizability can be seen as a limitation of our work. Indeed, we tested our hypotheses in Ukraine in the domain of an ongoing structural reform. Would the hypothesized relationships be the same in other countries? Would the effects remain when the process change has achieved its goals and the new system has become established? We do not know the answer yet. We see, however, that health care reforms are run regularly all over the world. Thus, the results might be of great interest; however, also the local conditions, for instance the financial and political situation, have to be taken into account.

8 Conclusions

With this work, we propose to integrate the JD-R model into BPM. Such integration allows for the inclusion of the physiological and psychological resources of individuals and, thus, helps make BPM human-centric. We developed a model with a set of direct and indirect effects of human-related constructs and showed that they are statistically significant for BPM-related aspects such as time (time span of diagnosing) and quality (errors in diagnosing). Openness to change showed a direct and partially mediated effect on cooperation issues of family doctors; burnout had a direct and mediated effect on errors in diagnosing; and, finally, cooperation issues had a direct and mediated effect on time span of diagnosing. Our paper suggests that appropriate process management would allow for the balancing of physiological and psychological job demands and resources.

References

1. Fuchs, V.R.: Economics, values, and health care reform. J. Am. Stat. Assoc. **86**(1), 1–24 (1996)
2. Saltman, R.B., Figueras, J.: European health care reform: analysis of current strategies. WHO Regional Publications European Series No. 72, pp. 5–38 (1997)
3. Dillon, K., Prokesch, S.: Megatrends in global health care. Harv. Bus. Rev. (2011)
4. Franco, L.M., Bennett, S., Kanfer, R.: Health sector reform and public sector health worker motivation: a conceptual framework. Soc. Sci. Med. **54**(8), 1255–1266 (2002)
5. Schaufeli, W.B.: Applying the job demands-resources model. Org. Dyn. **2**(46), 120–132 (2017)
6. WHO: Ukraine Reform Monitor, October 2017. http://carnegieendowment.org/2017/10/10/ukraine-reform-monitor-october-2017-pub-73330. Accessed 5 Jan 2018
7. OECD: European Commission (ed.): Oslo Manual: Guidelines for Collecting and Interpreting Innovation Data, 3rd edn, Paris (2005)
8. Haggerty, J.L., Reid, R.J., Freeman, G.K., Starfield, B.H., Adair, C.E., McKendry, R.: Continuity of care: a multidisciplinary review. BMJ **327**(7425), 1219 (2003)
9. Kvamme, O.J., Olesen, F., Samuelsson, M.: Improving the interface between primary and secondary care: a statement from the European working party on quality in family practice (EQuiP). Qual. Saf. Health Care **10**(1), 33–39 (2001)
10. Gérvas, J., Ferna, M.P., Starfield, B.H.: Primary care, financing and gatekeeping in Western Europe. Fam. Pract. **11**(3), 307–317 (1994)
11. Robinson, P.H.: Review article: recognition and treatment of eating disorders in primary and secondary care. Aliment. Pharmacol. Ther. **14**(4), 367–377 (2000)
12. Barkham, M., Gilbert, N., Connell, J., Marshall, C., Twigg, E.: Suitability and utility of the CORE-OM and CORE-A for assessing severity of presenting problems in psychological therapy services based in primary and secondary care settings. Br. J. Psychiatry **186**(3), 239–246 (2005)
13. Thompson, C., et al.: Effects of a clinical-practice guideline and practice-based education on detection and outcome of depression in primary care: Hampshire depression project randomised controlled trial. Lancet **355**(9199), 185–191 (2000)
14. Browne, G., et al.: Economic evaluations of community-based care: lessons from twelve studies in Ontario. J. Eval. Clin. Pract. **5**(4), 367–385 (1999)
15. Starfield, B.: Is strong primary care good for health outcomes. In: The Future of Primary Care: Papers for a Symposium Held on 13th September 1995. Office of Health Economics, pp. 18–29 (1995)
16. Barney, J.: Firm resources and sustained competitive advantage. J. Manag. **17**(1), 99–120 (1991)
17. Teece, D.J., Pisano, G., Shuen, A.: Dynamic capabilities and strategic management. Strateg. Manag. J. **18**(7), 509–533 (1997)
18. Hammer, M., Champy, J.: Reengineering the Corporation. HarperCollins, New York (1993)
19. vom Brocke, J., Rosemann, M.: Handbook on Business Process Management. Springer, Heidelberg (2010). https://doi.org/10.1007/978-3-642-01982-1

20. Dumas, M., Rosa, M.L., Mendling, J., Reijers, H.A.: Fundamentals of Business Process Management. Springer, Heidelberg (2013)
21. Kettinger, W.J., Teng, J.T., Guha, S.: Business process change: a study of methodologies, techniques, and tools. MIS Q. 55–80 (1997)
22. Wanberg, C.R., Banas, J.T.: Predictors and outcomes of openness to changes in a reorganizing workplace. J. Appl. Psychol. **85**(1), 132 (2000)
23. Schmelzer, H.J., Sesselmann, W.: Geschäftsprozessmanagement in der Praxis: Kunden zufriedenstellen, Produktivität steigern, Wert erhöhen, 8th edn. Hanser, Munich (2013)
24. Bakker, A.B., Demerouti, E., De Boer, E., Schaufeli, W.B.: Job demands and job resources as predictors of absence duration and frequency. J. Vocat. Behav. **62**(2), 341–356 (2003)
25. Demerouti, E., Bakker, A.B., Nachreiner, F., Schaufeli, W.B.: The job demands-resources model of burnout. J. Appl. Psychol. **86**(3), 499 (2001)
26. Bakker, A.B., Demerouti, E.: The job demands-resources model: state of the art. J. Manag. Psychol. **22**(3), 309–328 (2007)
27. Brotheridge, C.M., Grandey, A.A.: Emotional labor and burnout: comparing two perspectives of "people work". J. Vocat. Behav. **60**(1), 17–39 (2002)
28. Hobfoll, S.E.: The influence of culture, community, and the nested-self in the stress process: advancing conservation of resources theory. Appl. Psychol. **50**(3), 337–421 (2001)
29. Edwards, N., Saltman, R.B.: Re-thinking barriers to organizational change in public hospitals. Isr. J. Health Policy Res. **6**(1), 1–11 (2017)
30. Ashton-James, C.E., Ashkanasy, N.M.: Affective events theory: a strategic perspective. Res. Emot. Organ. (4), 1–34 (2008)
31. Dahl, M.S.: Organizational change and employee stress. Manag. Sci. **57**(2), 240–256 (2011)
32. Haines, V.A., Hurlbert, J.S., Zimmer, C.: Occupational stress, social support, and the buffer hypothesis. Work Occup. **18**(2), 212–235 (1991)
33. van Dam, K., Oreg, S., Schyns, B.: Daily work contexts and resistance to organisational change: the role of leader–member exchange, development climate, and change process characteristics. Appl. Psychol. **57**(2), 313–334 (2008)
34. Allen, J., Jimmieson, N.L., Bordia, P., Irmer, B.E.: Uncertainty during organizational change: managing perceptions through communication. J. Chang. Manag. **7**(2), 187–210 (2007)
35. Kristensen, T.S., Borritz, M., Villadsen, E., Christensen, K.B.: The Copenhagen burnout inventory: a new tool for the assessment of burnout. Work Stress **19**(3), 192–207 (2005)
36. Kowal, J., Gurba, A.: Mobbing and burnout in emerging knowledge economies: an exploratory study in Poland. In: Proceedings of the 49th Hawaii International Conference on System Sciences (HICSS), pp. 4123–4132 (2016)
37. Reichl, C., Leiter, M.P., Spinath, F.M.: Work–nonwork conflict and burnout: a meta-analysis. Hum. Relat. **67**(8), 979–1005 (2014)
38. Hair, J.F., Black, W.C., Babin, B.J., Anderson, R.E. (eds.): Multivariate Data Analysis, 7th edn. Prentice Hall, Upper Saddle River (2010)
39. Churchill, G.A.: A paradigm for developing better measures of marketing constructs. J. Mark. Res. **16**(1), 64–73 (1979)
40. Hu, L., Bentler, P.M.: Cutoff criteria for fit indexes in covariance structure analysis: conventional criteria versus new alternatives. Struct. Equ. Model. **6**(1), 1–55 (1999)
41. Fornell, C., Bookstein, F.L.: Two structural equation models: LISREL and PLS applied to consumer exit-voice theory. J. Mark. Res. **19**, 440–452 (1982)
42. Hair, J.F., Hult, G.T.M., Ringle, C.M., Sarstedt, M., Thiele, K.O.: Mirror, mirror on the wall: a comparative evaluation of composite-based structural equation modeling methods. J. Acad. Mark. Sci. **45**(5), 616–632 (2017)

43. Lee, R.T., Lovell, B.L., Brotheridge, C.M.: Tenderness and steadiness: relating job and interpersonal demands and resources with burnout and physical symptoms of stress in Canadian physicians. J. Appl. Soc. Psychol. **40**(9), 2319–2342 (2010)
44. Lee, R.T., Lovell, B.L., Brotheridge, C.M.: Relating physician emotional expression to shared understanding and shared decision-making with patients. Int. J. Work. Organ. Emot. **3**(4), 336–350 (2010)
45. Cherniss, C.: Staff Burnout: Job Stress in the Human Services. Sage Publications, Beverly Hills (1980)
46. Sobel, M.E.: Asymptotic confidence intervals for indirect effects in structural equation models. Sociol. Methodol. (13), 290–312 (1982)
47. MacKinnon, D.P., Fairchild, A.J., Fritz, M.S.: Mediation analysis. Annu. Rev. Psychol. **58**, 593–614 (2007)
48. Chin, W.W.: Commentary: issues and opinion on structural equation modeling. MIS Q. **22** (1), 7–16 (1998)
49. Grau, C., Moormann, J.: Empirical evidence for the impact of organizational culture on process quality. In: Proceedings of the 22nd European Conference on Information Systems (ECIS), Tel Aviv (2014)
50. Trkman, P.: The critical success factors of business process management. Int. J. Inf. Manag. **30**(2), 125–134 (2010)
51. Jeston, J., Nelis, J.: Business Process Management, 3rd edn. Routledge, London (2014)
52. Markos, S.M., Sridevi, M.S.: Employee engagement: the key to improving performance. Int. J. Bus. Manag. **5**(12), 89 (2010)
53. Burke, R.J., Greenglass, E.R., Schwarzer, R.: Predicting teacher burnout over time: effects of work stress, social support, and self-doubts on burnout and its consequences. Anxiety Stress Coping **9**(3), 261–275 (1996)

Process Performance Measurement System Characteristics: An Empirically Validated Framework

A. W. J. C. Abeygunasekera[1]([⊠]), Wasana Bandara[2], Moe T. Wynn[2], and Ogan Yigitbasioglu[2]

[1] University of Colombo, Colombo, Sri Lanka
janitha.abeygunsekera@hdr.qut.edu.au
[2] Queensland University of Technology, Brisbane, Australia
{w.bandara,m.wynn,ogan.yigitbasioglu}@qut.edu.au

Abstract. With the proliferation of business process reforms in organizations, the need for process-centric performance measurement systems is discussed in several studies. However, although there is considerable research on Performance Measurement Systems (PMSs) in general business contexts, research on performance measurement in process-centric contexts (e.g. within the BPM field) is scarce, and the characteristics of such process performance measurement systems (PPMSs) still remain poorly conceptualized, hindering their design and implementation. PPMSs are different from traditional performance measurement systems as the data gathered, and the information generated and disseminated are focused on processes, rather than on functions. This paper presents a PPMS characteristics framework, resulting from a multi-staged study design. Initially 38 PPMS characteristics were identified through a structured literature review which were then re-specified and confirmed with two in-depth case studies. The study findings resulted in an empirically supported PPMS characteristics framework, consisting of 30 PPMS characteristics grouped within 7 core themes, namely; (1) Quality characteristics; (2) Measurement Scope; (3) Contextual features considered in designing PPMS; (4) Relationship to organizational systems and structures; (5) Efficiency of information gathering and use; (6) Feedback and reporting; and (7) Potential uses of feedback generated. This is the first evidence-based synthesis of PPMSs characteristics and it provides a clear conceptualization and an understanding of what aspects a PPMS should comprise. The resulting framework will assist practitioners in designing and redesigning measurement systems in process centric contexts, which could in turn better support processes such as identifying improvement needs, measuring improved processes, benchmarking and controlling the processes, process maturity assessments, and benefits realization. The findings will also facilitate future research on PPMSs.

Keywords: Process performance measurement · Performance measurement · Business process management · Characteristics · Case study research

T. Hildebrandt et al. (Eds.): BPM 2019, LNBIP 360, pp. 211–227, 2019.
https://doi.org/10.1007/978-3-030-26643-1_13

1   Introduction

Performance measurement (PM) has been recognized as significant in many fields and process performance measurement is a well-established critical success factor for business process management (BPM) [1–3]. A process performance measurement system (PPMS) is expected to provide a systematic measurement of business processes [4] and it is a system that gathers process performance relevant data; compares the performance related values with other historical and target results/values and disseminates the results to the relevant process actors [5].

As stated by Tregear [6] "*if you don't measure process performance, you can't do process management - and you won't know if you are doing process improvement*" (p. 5). The 'measurement' aspect of process performance has therefore gained attention both in BPM practice [7] and as a significant and growing research area [8, 9]. It is important for the measures to be 'process-centric' in order to mitigate the risks faced in achieving the expected results of the process changes [10].

Substantial benefits can be gained in strategic and operational management through the use of effective process performance measures [11]. PPMSs provide evidence for the value/impact of efficient and effective processes, enabling these improved processes to be better embedded in daily operations [12]. The BPM literature recognizes the importance of process-centric performance measurements to enable better BPM adoption outcomes [4, 12]. However, organizations tend to adopt different elements of PPMSs, and measure processes 'ad-hoc' and not continuously [13, p. 615], and the measurement of the impact of process improvements is often lacking [e.g. 4, 5, 8, 9]. Moreover, process improvements are more likely to have a positive impact on overall organizational performance when the performance measurement system (PMS) is designed or redesigned and deployed appropriately [14]. Likewise, inadequate measurement generates uncertainties regarding the impact of process improvement initiatives [15]. Thus, a well-designed process performance measurement system is needed for planning, designing and controlling business processes from the early stages of the process change/improvement initiatives [16, 17].

Traditional performance measures that focus on individual organizational functions do not cater well to contemporary business needs [18, 19] and the measurement needs of process improvement initiatives [20]. How to construct such process focused performance measures that do cater to the measurement needs of process improvement initiatives is not yet well understood. While there is considerable research on PMSs in general business contexts, research on performance measurement in process-centric contexts (e.g. within the BPM field) is scarce. For any field to develop and be more relevant to theory and practice, the characteristics of a system under consideration need to be made specific and explicit [21]. Explication of characteristics also facilitates consistent measurement and effective comparisons with other PMSs [22]. Thus, to develop and deploy relevant PPMSs or to redesign existing PMSs to better suit process-centric contexts, it is essential to identify the required characteristics of a PPMS; but such is not available in the current body of knowledge.

The research question driving this study is: **What are the characteristics of a process performance measurement system?** The aim of this work is to derive a comprehensive list of PPMS characteristics which are essential to consider when designing and deploying a PPMS. The framework with its evidence supported structured set of characteristics, can serve as a useful reference point for PPMS practice and future research. In the context of this study, 'characteristics' are the key features (or attributes) that form and describe a PPMS (i.e. the 'what' aspects). Note that we do not consider the PPMS design-steps (the 'how' aspects) when aiming to characterize PPMSs. While the PPMS characteristics could have specific design implications across the diverse stages of developing a PPMS, the focus of this paper is solely on deriving the PPMS characteristics (the design guidelines will be part of future research that follows).

A critical synthesis of the existing literature on PPMS characteristics is presented next (Sect. 2), to further justify the motivation for this work. The study comprised a multi-staged design as outlined in Sect. 3, with a structured literature review (SLR) (Sect. 3.1) followed by two in-depth case studies (Sect. 3.2). The study findings resulted in an empirically supported PPMS characteristics framework, consisting of 30 PPMS characteristics grouped within 7 core themes (Sect. 4).

2 Overview of the Existing Literature on PPMS Characteristics

A review of the literature on performance measurements in process-centric contexts confirms the limited research on PPMS characteristics. However, considerable research relating to other aspects of PMSs in process-centric contexts does exist. For instance, many scholars discuss the need for the adaptation of the traditional PMSs to BPM contexts [5, 6, 9, 12, 23, 24] including the need for methods to depict the financial/economic consequences of the process improvement efforts [17, 25, 26] as well as the importance of measuring process orientation [27]. Some refer to *managerial issues*, such as the role of process performance measurement in BPM adoption outcomes [4]; example cases of benefits obtained by tracking process performance [28]; the value of PPMSs [29]; factors influencing the ineffectiveness of PPMSs [29]; process performance measurement as a component of maturity assessment [30, 31]; process owner role and process performance measurement [32]; having a strategic approach within the organization for BPM and its impact on process performance measurement [11]. Others present certain *PPMS attributes* - defining process performance indicators [33]; review on the current state of research on PPMSs [34] and specific PPMS indicators and metrics [35].

Even though the need for a clearer conceptualization of PPMSs is often mentioned, detailed discussions on PPMS characteristics are still very limited. A systematic literature search (see Sect. 3) conducted on PPMSs revealed only eight papers that proposed potential characteristics for process-centric contexts. Three papers concerned PPMS characteristics in general and the other five discussed performance measurement characteristics in specific process improvement contexts; such as Just-in-Time (JIT) and Total Quality Management (TQM) environments.

With regard to the three papers focusing specifically on PPMS characteristics, Kueng and Krahn [24, p. 153] mention only one characteristic. i.e. *"PPMS should present an integral and holistic view of the performance of business processes*". Kueng [5] suggests three characteristics for a PPMS. It should be: focused on processes; evaluate performance holistically; and designate responsible parties (to each indicator to be measured). As suggested by Wieland et al. [36], this paper also argues that these characteristics are incomplete and too vague to operationalize. Wieland et al. [36] discuss PPMS characteristics essential for successful customer-orientation and identified customer demands and design features of PPMSs through a literature review. The limitations of the Wieland et al. [36] framework are; the 'customer only' focus, and how they mix the characteristics and design-steps to be considered when designing a PPMS etc. (as presenting them all together does not assist in the delineation of the precise characteristics of a fully developed PPMS).

The other five papers discussed PMS but did not focus on PPMS explicitly. Two papers discussed the characteristics of a PMS in a Just-in-Time environment [37, 38]. Schalkwyk [39] discussed what the PMS should consist of, within a TQM environment. The use of the PMS framework to enable and guide sound process re-engineering endeavours in maintenance was presented by Kutucuoglu et al. [40]. Bond [41] discussed PMS within a kaizen and re-engineering context and proposed that the characteristics of the different stages of process life cycles need to be considered when determining the performance metrics and when designing approaches for monitoring and controlling the processes. All the characteristics proposed in these eight papers are considered in our analysis and included within the preliminary set of 38 characteristics.

It was evident from the review that the PPMS literature to date is lacking a comprehensive synthesis. We also noted that the existing discourses often only addressed specific process-centric aspects and did not cover some of the essential characteristics that should be found in any performance measurement system (e.g. controls to minimize the opportunity for manipulation of the measures and results etc.). With the underlying assumption that the PPMS is a type of PMS, we argue that for a characterization of PPMS to be complete, it should not only focus on process centric specificities, but also have all essential features that any PMS should have. We maintained this view in our research design and the development of our framework.

3   Methodology

This study followed a multi-phased approach, using a structured literature review to identify potential PPMS characteristics and two in-depth case studies to provide empirical validation. The phases are summarized in Sects. 3.1 and 3.2.

3.1   A Structured Literature Review (SLR) to Synthesize PPMS Characteristics

The authors aimed to collate and synthesize the scattered knowledge on the characteristics of PPMSs (for the specific process-centric features) and characteristics of PMSs in literature (based on the assumption that PPMS is a type of PMS and hence should include the basic features of a PMS adapted to a process-centric context). The SLR applied in this study followed the guidelines proposed by Bandara et al. [42].

After a background search on the topic areas, the main keywords - 'performance measurement system' and 'process performance measurement system' were selected. A search string combining these two terms was used to search in ABI/INFORM, EBSCOhost and Emerald Insight in August 2017, resulting in 665 potentially relevant papers. A relevance check was carried out by two of the authors at several iterations with the primary rule being; 'the paper must *directly* discuss characteristics of PMSs or PPMSs'. Conceptual papers (i.e. those without supporting empirical evidence) were included for completeness purposes. 38 papers resulted from this relevancy screening. Most of the papers were removed from the analysis as they either discussed other aspects of PMS/PPMS such as measurement system design stages or did not discuss the characteristics in a collective manner. Backward and forward searches were carried out to identify further relevant papers, and 15 peer-reviewed papers and 3 book chapters were added. The 3 book chapters were not peer-reviewed but were included as many other papers referred to them and it is recognized that this sort of 'grey literature' can add value, especially to under-researched areas [43].

The 56 sources[1] were then exposed to a detailed coding protocol- with coding guidelines [44] that were specifically designed and pre-tested. PMS/PPMS characteristics were inductively extracted and recorded using an Excel Spreadsheet [42], resulting in a total of 418 open codes. These were tabulated separately as PMSs and PPMSs characteristics. Iterative reviews by the research team of these open codes resulted in some of the characteristics being broken down into further elements (to represent atomic sub-themes) and some (those that were very similar) being merged together, resulting in 473 total re-specified open codes. In the next round, axial coding was applied by forming coding families (following [45, 46]); resulting in the open codes being synthesized into 38 discrete PPMS characteristics (C1–C38). Appendix A lists and describes the 38 PPMS characteristics, and shows the sources supporting each. These characteristics were validated through two case studies as described in the next section.

3.2 In-depth Case Studies to Validate a-Priori PPMS Characteristics

Multiple-case designs are desirable when the intent of the research is description, theory building, or theory testing [47] and multiple case studies enable researchers to explore differences within and between cases and to replicate the findings across cases [48]. Thus, two case studies are conducted to validate the PPMS characteristics instead of a single case study.

The cases were selected based on a pre-defined case selection criterion (recommended in other studies (i.e. [49, 50])). A qualifying case had to be an improvement initiative implemented in a core business process in an organization that had implemented multiple process improvement initiatives and have a PPMS and/or PMS in place within the organization). Two cases in two large scale multinational manufacturers in Sri Lanka were studied. One was an apparel manufacturer (subsequently

---

[1] See References provided in Appendix A.

referred to as AMC) and the other was a tyre manufacturer (subsequently referred to as TCC). The unit of analysis was the process improvement initiative).

In AMC, the process improvement initiative concerned an improvement to one of their core business processes – 'sewing process' was studied in depth. AMC is the first apparel manufacturer in Sri Lanka, to introduce this kind of a process improvement initiative (subsequently referred to as the 'Dancing module'). They planned and tested it for one and a half years. Implementation was carried out in stages, where firstly 60 team members who worked in 2 lines were divided into 3 lines of 20 members each to increase productivity. Thereafter all the lines in the plant were converted. The team members were the lowest level employees in the plant and prior to the dancing module, they were always seated when sewing the garments. With the change, they stand and stitch the whole day and rotate to at least 3 machines. Most of the team members were only exposed to the seated mode of working, in this 30-year-old manufacturing plant. Thus, introducing the dancing module was a major transformation.

In TCC, the unit of analysis studied was an improvement to the mould operation (subsequently referred to as 'change in mould operation') - one of the core and critical processes in their tyre manufacturing process. TCC is the largest multinational tyre manufacturer in Sri Lanka with eleven plants. This was a unique case, in the following regard; that the initiative took place after a set of failed initiatives introduced by the Centre of Excellence for BPM in the organization, together with external consultants. According to the study respondents, the main reason for the failure in the earlier initiatives had been the lack of properly aligned measures. They planned and tested the new change initiative between 6–12 months in one plant prior to implementing to the full factory floor, which was implemented next within three plants and later expanded to the other eight plants.

Interviews and document analysis were used as the primary data collection mechanisms. Semi-structured interview protocols were used to guide the interview process which was conducted in two stages; for primary data collection, and for results confirmation. 12 interviews were conducted at AMC (*with - Manager Lean Enterprise - BPM Centre of excellence, Deputy General Manager – HR, Group Head of Finance, Management Accountant, Operations Manager, Production Control Unit Manager, Head/Manager Lean Enterprise, Project Champion cum Industrial Engineering Executive, Lean Implementation Executive and three Sewing Machine Operators*) and 10 interviews at TCC; *with - Program Manager (Strategic Supply Chain Process Improvements), Plant Director, Industrial Engineering Senior Manager, Product and Process Senior Engineer, Product and Process, IE & Lean Senior Executive Engineer and two Shift in-charges*). The duration of interviews varied between 30–90 min, and some respondents were approached twice for further clarifications. See Table B.1 of Appendix B for further details. NVivo tool was used as a data/evidence management tool for managing and coding the interview transcripts. A coding rule book and a pre-defined node structure were used for the analysis. For instance, a statement such as: "*The measures will be defined in detail very clearly so that we know what we are measuring and how and for what objective they will be introduced*" is coded under C1-Performance measures being clearly defined, with an explicit purpose. See Appendix B, for detailed case evidence pertaining to the 38 characteristics. A summary of case study results is presented next.

4 Summary Case Study Results

The a-priori 38 PPMS characteristics derived from the literature review (see Sect. 3.1 and Appendix A) were re-specified and confirmed through two case studies and adjusted (based on case evidence) to derive the final PPMS characteristics framework. Two new PPMS characteristics were identified and added to the list of characteristics (Sect. 4.1), and some characteristics with weak supporting evidence were still taken forward (Sect. 4.2), and some characteristics were merged due to clear similarities observed through the case data (Sect. 4.3). They were then categorized into the seven core themes as discussed in Fig. 1 in Sect. 5.

4.1 Newly Added Characteristics

Two characteristics that were not covered in the initial 38 characteristics (see C1–C38 Appendix A) were added and are explained further below.

(New-1) - Performance measures should provide the ability to challenge the jobs and encourage employees to work at a higher scale. [This was taken as a characteristic due to statements such as: *"our PMS is always challenging the job roles. We try to make the job challenging to employees so that they will stretch towards higher targets. That is an expectation from the employees too"* - Production Control Unit Manager (AMC) and *"in the first six months we didn't have proper measurements/KPIs, it was only basic things like monitoring within 24 h. We tightened the KPIs step by step ... They did the same work, but the measures were getting tight and they had to work at a higher speed"* - Industrial Engineering Manager (TCC)].

(New-2) - Performance measurements should be able to guide the employees on how to perform the expected tasks and to what extent they should be performed. [This was taken as a characteristic due to statements such as: *"people should know the guideline to perform. More than pushing the people, KPIs are like guidelines for them. When you take a job description it is very broad. It doesn't show the KPIs. This is important. When you work you should know what the goals are. This is guidelines for them to work and also for us to have as a rewards base"* - Deputy General Manager - HR (AMC) and *"together we define the KPIs and they get the index cards and they will make their staff work on them. Then they know where they should reach and what they need to reach there"* - Plant Director (TCC)].

4.2 A-Priori Characteristics with Conflicting Empirical Support

While the a-priori 38 characteristics were instantiated across the two cases (see Appendix B for supporting case evidence), two (C6 and C15) were subject to conflicting views and were only weakly confirmed; but were still included in the final list.

C6 (The set of performance measures should be few but complete and critical). C6 was not instantiated with evidence from current practice by both cases. But, AMC respondents believed that it is important to have few measures and they were attempting to reduce the number of measures and KPIs. TCC respondents did not believe in having few measures but did believe in having the most important set of

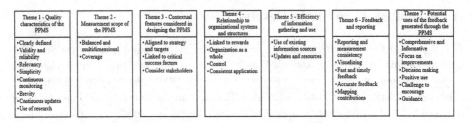

Theme 1 - Quality characteristics of the PPMS	Theme 2 - Measurement scope of the PPMS	Theme 3 - Contextual features considered in designing the PPMS	Theme 4 - Relationship to organizational systems and structures	Theme 5 - Efficiency of information gathering and use	Theme 6 - Feedback and reporting	Theme 7 - Potential uses of the feedback generated through the PPMS
• Clearly defined • Validity and reliability • Relevancy • Simplicity • Continuous monitoring • Brevity • Continuous updates • Use of research	• Balanced and multidimensional • Coverage	• Aligned to strategy and targets • Linked to critical success factors • Consider stakeholders	• Linked to rewards • Organization as a whole • Control • Consistent application	• Use of existing information sources • Updates and resources	• Reporting and measurement consistency • Visualizing • Fast and timely feedback • Accurate feedback • Mapping contributions	• Comprehensive and Informative • Focus on improvements • Decision making • Positive use • Challenge to encourage • Guidance

Fig. 1. The resulting PPMS Characteristics framework.

measures to cover all important aspects that need to be measured for decision making related to the process improvement initiatives. With C15 (Linked to critical success factors and key business drivers) the respondents of both cases saw value in this characteristic but stated that it is not consistently required for all process improvement initiatives. Critical success factors (CSFs) are identified and measured only when a process improvement initiative is costly and/or complex. Some CSFs were considered in the AMC initiative studied but none were considered in the TCC initiative.

4.3   Merging of Characteristics

Certain characteristics were deemed best to be merged as the respondents were attributing the same meaning to them, even though they were discussed separately in the literature. Some were further refined to better represent the attributed meaning. A summary of the characteristics that were merged after the analysis is presented below.

- C3 [Performance measures being relevant to the business process, the people who are accountable for the process; and the output of the process]. '*Relevant to the business process*' has a similar meaning to C19 - [performance measures integrated with process execution and connected to the KPI and the process steps]. Therefore, C19 was absorbed into C3.
- C9 [All-inclusive/balanced/multi-dimensional set of measures] and C11 [Use of Both Financial, non-financial/Objective, subjective/quantitative, qualitative measures] were discussed by the respondents as having similar meanings.
- C10 [Performance measures should take both long-term and short-term views into account] and C17 [Performance measures should track the past and present performance and performance that influences on future activities/performance] were merged as the 'present' and 'short-term', and the 'future' and 'long-term' were taken by respondents to mean the same.
- C12 [Use of trend and ratio-based performance measures] and C28 [Visuals can make more impact than numbers] are merged because the respondents mentioned that they visualize through ratios and trend lines etc. in graphs.
- C13 [Performance measures should change dynamically and be consistent or coherent with the organizational strategy to support strategy realization] and C14 [Performance measures linked to targets, goals, and objectives] were discussed as meaning the same by respondents.

- C20 [Performance measures consider the organization *as a whole*, to minimize conflict] and C21 [Focus on processes and *integration of functions*]; especially the 'Integration of functions' in C21, was deemed similar to considering 'measures as a whole' in C20.
- C25 [Performance measures should be cost effective to use] was considered a sub part of C24 [Use of automatically collected data and the existing sources of data], as further developments to PPMSs were conducted with the expectation of getting higher benefits from the PPMS in future as per the respondents.
- In C31 [Reports should be made in a simple, *frequent and regular manner*, available constantly for review/use, and not used as a replacement for review meetings]. The aspect of 'Not used as a replacement for review meetings' was not mentioned by respondents and they mentioned that they always review the generated reports. Therefore, C31 was adjusted by adding 'reviewed' to it – [Reports should be made in a simple, frequent and regular manner, available constantly for review/use, and be reviewed regularly].

Further C27 [Maintain *consistency* over time and in *reporting*] and C31 were merged, as 'consistency' in 'reporting' in C27 was deemed similar in meaning by respondents to 'reports should be made in a … frequent and regular manner' in C31.

- C33 [...provide comprehensive information to users with easily identifiable and useful relationships...] and C36 [...provide information for actionable results. Remedial action ...] were merged as respondents did not differentiate between these two ideas. Therefore, C36 was positioned as a subpart of C33.
- C34 [Performance measurement results - Enable consistent benchmarking/comparisons...] and C35 [...focus on improvements and inspire and permit employees to monitor, control and further improve ...] were merged, as many of the respondents indicated that continuous improvement had been done by benchmarking and through comparisons. Therefore, C34 was positioned as a sub-part of C35.

5   PPMS Characteristics Framework and Summary Discussion

With the additions and mergers, 30 PPMS characteristics were derived as the confirmed set of PPMS characteristics. Codes for the confirmed characteristics (*column 3*), codes for the original and merged characteristics (*column 1*) together with the number of literature sources for each of the 38 original characteristics (*column 2*) and sources excluding overlaps related to merged characteristics (*column 4*) and the finalized characteristics (*column 5*) are presented in Table 1.

Table 1. The confirmed 30 PPMS characteristics.

1	2	3	4	5	6	7	8
					PMS	PPMS	PMS & PPMS
Related Old Characteristic(s)	Literature Sources before merging	New Codes of the confirmed Characteristics (NC)	Literature Sources after merging	Confirmed Characteristics			
C1	11	NC1	11	Clearly defined			X
C2	7	NC2	7	Validity and reliability			X
C3	9	NC3	**13**	Relevancy			**X**
C19	6						
C4	23	NC4	23	Simplicity			X
C5	1	NC5	1	Continuous monitoring		X	
C6	7	NC6	7	Brevity	X		
C7	16	NC7	16	Continuous updates			X
C8	1	NC8	1	Use of research	X		
C9	16	NC9	**35**	Balanced and multi-dimensional			**X**
C11	28						
C10	6	NC10	**11**	Coverage	X		
C17	5						
C13	24	NC11	**34**	Aligned to strategy and targets			**X**
C14	17						
C15	8	NC12	8	Linked to critical success factors	X		
C16	17	NC13	17	Consider stakeholders			X
C18	7	NC14	7	Linked to rewards	X		
C20	6	NC15	**16**	Organization as a whole			**X**
C21	11						
C22	13	NC16	13	Control			X
C23	17	NC17	17	Consistent application			X
C24	6	NC18	**10**	Use of existing information sources			**X**
C25	4						

1	2	3	4	5	6	7	8
					PMS	PPMS	PMS & PPMS
Related Old Characteristic(s)	Literature Sources before merging	New Codes of the confirmed Characteristics (NC)	Literature Sources after merging	Confirmed Characteristics			
C26	4	NC19	4	Updates and resources			X
C27	3			Reporting and measurement consistency			X
C31	4	NC20	6				
C28	4	NC21	9	Visualizing			X
C12	7						
C29	11	NC22	11	Fast and timely feedback	X		
C30	9	NC23	9	Accurate feed-back			X
C32	10	NC24	10	Mapping contri-butions			X
C33	11	NC25	15	Comprehensive and informative			X
C36	6						
C34	10	NC26	24	Focused on improvements			X
C35	18						
C37	12	NC27	12	Decision mak-ing			X
C38	4	NC28	4	Positive use			X
Newly added		NC29	Not applicable (NA)	Challenge to encourage	NA	NA	NA
Newly added		NC30	NA	Guidance	NA	NA	NA

Using the literature review and case data as the evidence base, these 30 PPMS characteristics were inductively analyzed (through multiple iterations supported by multi-coder corroboration sessions) to identify higher-level themes in which to group them. This resulted in 7 themes, summarized in Fig. 1:

1. Quality characteristics of the PPMS: the designers of PPMSs should take these quality characteristics into account as they influence efficiency and effectiveness, to assure that the PPMS is of high quality.

2. Measurement scope of the PPMS: are the characteristics that ensure the provision of an overall assessment of the performance of the processes within an organization. The measurements should include financial and non-financial measures, as well as short and long term measures.
3. Contextual features considered in designing the PPMS: refers to characteristics relating to critical success factors, key business drivers, the organizational strategy, targets and goals, and the stakeholder needs.
4. Relationship to organizational systems and structures: characteristics regarding the importance of linking PPMSs to other systems and structures in the organization such as linking PPMS to rewards systems.
5. Efficiency of information gathering and use: characteristics that stress the need for PPMSs to be cost effective while using the existing sources of data and formally training the employees in the use of PPMS.
6. Feedback and reporting: characteristics of how the PPMS should conduct reporting and provide feedback. Consistency and transparency of generated information and frequent reporting cycles are emphasized.
7. Potential uses of the feedback generated through the PPMS: The list is not exhaustive but shows the characteristics referring to potential use cases of PPMSs.

The PPMS Characteristics (NC1–NC30 – see Table 1 for further supporting detail) mapped against these seven themes are presented in Fig. 1, forming the final PPMS Characteristics framework. Even though some characteristics were mentioned only in relation to PMS (e.g. C6, C8 etc.) or PPMS (C5), most of the others were mentioned related to both PMS and PPMS in literature (these details are depicted in the last three columns in Table 1). But based on the respondents of the two case studies, we argue that all 30 characteristics should be embedded into a PPMS as they serve different purposes (indicated by the names of the themes into which they are grouped).

Measuring process performance is a precondition for analyzing and subsequently improving business processes [51], and process performance measurement is a critical phase of any process improvement lifecycle [52]. As described by Hernaus, [11] significant benefits can be gained in strategic and operational management through the use of effective process performance measures. However, having the 'right' measurement system and the supporting process-centric data has been a long-term challenge in the field [51], due to the lack of PPMSs. This lack of effective measurement systems and accurate and easily obtainable process data (past and present) hinders process improvement initiatives [53, 54], the ability to accurately identify pain points, and benchmark or show the impact of process improvements efforts.

The study results presented here provide the necessary evidence-based guidelines to address this gap of developing a PPMS, in order to measure process performance data by identifying the characteristics necessary for a PPMS. These characteristics become a useful checklist for practitioners designing and/or evaluating PPMSs. While recent developments in Business Process Management Systems (BPMSs), especially data-centric enhancements, have improved access to process-centric data, more holistic guidance on how such access to process-centric data should be designed and integrated within organizational contexts is still lacking. This empirically confirmed framework with the seven themes and 30 characteristics provides a holistic overview, complimenting such technical developments in BPMSs; it points to how process performance measurement should contribute towards quality assurance, cost management, integration to different systems and structures, and how process performance measures need to dynamically evolve within diverse contexts. It also captures how the feedback loops and reporting resulting from PPMS outcomes should be set up and how to best operationalize the insights obtained from deploying a PPMS. This framework guides organizations on how to move away from current 'ad-hoc' [13] process measurement efforts, reduce existing uncertainties [15], and (re-) design PPMSs appropriately [14] with a holistic and long-term view.

6 Conclusion

Conceptualizing PPMSs is a critical gap that is unaddressed, and this study attempts to fill that gap by presenting an empirically supported PPMS characteristics framework that identifies what characteristics a PPMS should contain. A multi-staged study design was followed where 38 PPMS characteristics identified through a structured literature review were re-specified and confirmed with two in-depth case studies resulting in 30 PPMS characteristics, which were then categorized into 7 themes. This empirically confirmed PPMS framework (the first of its kind) can serve as a useful reference point for PPMS practice and future research. Compared to the available studies on PMS/PPMS characteristics (within the identified fifty-six papers) this is a complete collection of characteristics synthesized to make the scattered knowledge on characteristics to a single place. The list of characteristics can be used as a complete check list for practitioners to design PPMSs and to review and adapt traditional PMSs. The collection and synthesis were done in a rigorous manner. The method followed in building the themes as well as the characteristics is tracked through Excel spread sheets and Nvivo data management tool. Thus, having a trail of evidence between initial extracts from the papers and meta themes. The identified list facilitates in making sense and have eased the attempts of taking actions on PPMS. Themes provide a clearer simplified and actionable set of areas that needs to be addressed when designing PPMSs or reviewing or adapting traditional PMSs to suit process-centric settings.

While a rigorous literature and case study-based approach was followed, the results may have some limitations. While a literature based a priori models provide guidance for empirical work, they can also influence 'what is sought for' in the empirical settings; hence introduce some bias. The framework may also be affected by usual limitations of qualitative research such as; selection- and researcher-bias, despite the protocols followed to minimize these. Furthermore, we acknowledge that the PPMS characteristics can be dynamic in nature (i.e. evolve and change in diverse contexts and over time). Given that the cases studied here were from the manufacturing sector, we recommend future research that investigates and validates the PPMS characteristics framework in other sectors. Future research can also focus on identifying the PPMS characteristics that are differently relevant to diverse process improvement project types [49]; and better understand which characteristics are essential for all project types and which are important for only some, as well as how they change in diverse contexts and over time. Future research can also investigate evolving PPMS design-steps (i.e.

Blasini et al. 2018), in particular to better understand when and how the identified PPMS characteristics should be integrated into the PPMS design process.

Appendix A

PPMS characteristics extracted from the Structured Literature Review, available at; http://www.workflowpatterns.com/janitha/Appendix%20A.pdf

Appendix B

Case study evidence supporting the changes made to the a priori PPMS Characteristics, available at; http://www.workflowpatterns.com/janitha/Appendix%20B.pdf

References

1. Bai, C., Sarkis, J.: A grey-based DEMATEL model for evaluating business process management critical success factors. Int. J. Prod. Econ. **146**(1), 281–292 (2013)
2. Alibabaei, A., Bandara, W., Aghdasi, M.: Means of achieving business process management success factors. In: Proceedings of the 4th Mediterranean conference on information systems. Department of Management Science & Technology, Athens University of Economics and Business (2009)
3. Trkman, P.: The critical success factors of business process management. Int. J. Inf. Manag. **30**(2), 125–134 (2010)
4. Vuksic, V.B., Glavan, L.M., Susa, D.: The role of process performance measurement in BPM adoption outcomes in Croatia. Econ. Bus. Rev. Central South - Eastern Europe **17**(1), 117–150 (2015)
5. Kueng, P.: Process performance measurement system: a tool to support process-based organizations. Total Qual. Manag. **11**(1), 67–85 (2000)
6. Tregear, R.: Putting Process at the Center of Business Management. BPTrends September 2013, pp. 1–13 (2013)
7. Minonne, C., Turner, G.: Business process management - are you ready for the future? Knowl. Process Manag. **19**(3), 111–120 (2012)
8. Abeygunasekera, A.W.J.C., Bandara, W., Wynn, M., Yigitbasioglu, O.: Nexus between Business Process Management (BPM) and accounting: a literature review and future research directions. Bus. Process Manag. J. **24**(3), 745–770 (2018)
9. Choong, K.K.: Are PMS meeting the measurement needs of BPM? A literature review. Bus. Process Manag. J. **19**(3), 535–574 (2013)
10. vom Brocke, J., Sonnenberg C.: Process Management and Accounting-An Overdue Take on Measuring the Economic Value of Business Processes. Class Notes: BPM and Education and Research (2014)
11. Hernaus, T., Bach, M.P., Vuksic, V.B.: Influence of strategic approach to BPM on financial and non-financial performance. Baltic J. Manag. **7**(4), 376–396 (2012)

12. Tregear, R.: Practical Process: Measuring Processes. BPTrends November 2012, pp. 1–9 (2012)
13. Vuksic, V.B., Bach, M.P., Popovic, A.: Supporting performance management with business process management and business intelligence: a case analysis of integration and orchestration. Int. J. Inf. Manag. **33**(4), 613–619 (2013)
14. Edson, P.D.L., Gouvea da Costa, S.E., Angelis, J.J., Munik, J.: Performance measurement systems: a consensual analysis of their roles. Int. J. Prod. Econ. **146**(2), 524–542 (2013)
15. Lacerda, R.T.D.O., Ensslin, L., Ensslin, S.R., Knoff, L., Martins, D.J.C.: Research opportunities in business process management and performance measurement from a constructivist view. Knowl. Process Manag. **23**(1), 18–30 (2016)
16. Kuwaiti, M.E., Kay, J.M.: The role of performance measurement in business process re-engineering. Int. J. Oper. Prod. Manag. **20**(12), 1411–1426 (2000)
17. Sonnenberg, C., vom Brocke, J.: The missing link between BPM and accounting. Bus. Process Manag. J. **20**(2), 213–246 (2014)
18. Bititci, U.S., Carrie, A.S., McDevitt, L.: Integrated performance measurement systems: a development guide. Int. J. Oper. Prod. Manag. **17**(5), 522–534 (1997)
19. Tung, A., Baird, K., Schoch, H.P.: Factors influencing the effectiveness of performance measurement systems. Int. J. Oper. Prod. Manag. **31**(12), 1287–1310 (2011)
20. Amaratunga, D., Baldry, D., Sarshar, M.: Process improvement through performance measurement: the balanced scorecard methodology. Work Study **50**(5), 179–188 (2001)
21. Franco-Santos, M., et al.: Towards a definition of a business performance measurement system. Int. J. Oper. Prod. Manag. **27**(8), 784–801 (2007)
22. Keathley, H., Du, R., Olliges, K.: Review of performance measurement practices in military and government sectors. In: Industrial and Systems Engineering Research Conference, Nashville, Tenn, pp. 1087–1096 (2015)
23. Beckley, G.B.: Metrics for continuous improvement: tapping the potential of legacy systems. J. Syst. Manag. **45**(12), 20–21 (1994)
24. Kueng, P., Krahn, A.J.W.: Building a process performance measurement system: some early experiences. J. Sci. Ind. Res. **58**(3), 149–159 (1999)
25. Housel, T.J., Morris, C.J., Westland, C.: Business process reengineering at Pacific Bell. Plan. Rev. **21**(3), 28–33 (1993)
26. vom Brocke, J., Recker, J., Mendling, J.: Value-oriented process modeling: integrating financial perspectives into business process re-design. Bus. Process Manag. J. **16**(2), 333–356 (2010)
27. Kohlbacher, M., Gruenwald, S.: Process orientation: conceptualization and measurement. Bus. Process Manag. J. **17**(2), 267–283 (2011)
28. Dawson, R., O'Neill, B.: Simple metrics for improving software process performance and capability: a case study. Softw. Qual. J. **11**(3), 243–258 (2003)
29. Nenadál, J.: Process performance measurement in manufacturing organizations. Int. J. Prod. Perform. Manag. **57**(6), 460–467 (2008)
30. Rohloff, M.: Case study and maturity model for business process management implementation. In: Dayal, U., Eder, J., Koehler, J., Reijers, H.A. (eds.) BPM 2009. LNCS, vol. 5701, pp. 128–142. Springer, Heidelberg (2009). https://doi.org/10.1007/978-3-642-03848-8_10
31. Rosemann, M., De Bruin T.: Application of a holistic model for determining BPM maturity. BPTrends February 2005, pp. 1–21 (2005)
32. Kohlbacher, M., Gruenwald, S.: Process ownership, process performance measurement and firm performance. Int. J. Prod. Perform. Manag. **60**(7), 709–720 (2011)

33. del-Río-Ortega, A., Resinas, M., Cabanillas, C., Ruiz-Cortés, A.: On the definition and design-time analysis of process performance indicators. Inf. Syst. **38**(4), 470–490 (2013)

34. Glavan, L.M.: Understanding process performance measurement systems. Bus. Syst. Res. **2**(2), 25–38 (2011)

35. Van Looy, A., Shafagatova, A.: Business process performance measurement: a structured literature review of indicators, measures and metrics. SpringerPlus **5**(1), 1797 (2016)

36. Wieland, U., Fischer, M., Pfitzner, M., Hilbert, A.: Process performance measurement system – towards a customer-oriented solution. Bus. Process Manag. J. **21**(2), 312–331 (2015)

37. Crawford, K.M., Cox, J.F.: Designing performance measurement systems for just-in-time operations. Int. J. Prod. Res. **28**(11), 2025–2036 (1990)

38. Lea, R., Parker, B.: The JIT spiral of continuous improvement. Ind. Manag. Data Syst. **89**(4), 10–13 (1989)

39. Schalkwyk, J.C.: Total quality management and the performance measurement barrier. TQM Mag. **10**(2), 124–131 (1998)

40. Kutucuoglu, K.Y., Hamali, J., Sharp, J.M., Irani, Z.: Enabling BPR in maintenance through a performance measurement system framework. Int. J. Flexible Manuf. Syst. **14**(1), 33–52 (2002)

41. Bond, T.C.: The role of performance measurement in continuous improvement. Int. J. Oper. Prod. Manag. **19**(12), 1318–1334 (1999)

42. Bandara, W., Furtmuller, E., Gorbacheva, E., Miskon, S., Beekhuyzen, J.: Achieving rigour in literature reviews: insights from qualitative data analysis and tool-support. Commun. Assoc. Inf. Syst. **37**(1), 154–204 (2015)

43. Adams, R.J., Smart, P., Huff, A.S.: Shades of grey: guidelines for working with the grey literature in systematic reviews for management and organizational studies. Int. J. Manag. Rev. **19**(4), 432–454 (2017)

44. Creswell, J.W.: Research Design: Qualitative, Quantitative, and Mixed Methods Approaches, 3rd edn. Sage Publications, Thousand Oaks (2009)

45. Corbin, J.M., Strauss, A.L.: Basics of Qualitative Research: Techniques and Procedures for Developing Grounded Theory, 3rd edn. SAGE Publications, Los Angeles (2008)

46. Wolfswinkel, J.F., Furtmueller, E., Wilderom, C.P.: Using grounded theory as a method for rigorously reviewing literature. Eur. J. Inf. Syst. **22**(1), 45–55 (2013)

47. Benbasat, I., Goldstein, D., Mead, M.: The case research strategy in studies of information systems. MIS Q. **11**(3), 369–386 (1987)

48. Yin, R.K.: Case Study Research: Design and Methods, 4th edn. Sage Publications, Thousand Oaks (2009)

49. vom Brocke, J., Zelt, S., Schmiedel, T.: Considering Context in Business Process Management: The BPM Context Framework. BPTrends December 2015, pp. 1–13 (2015)

50. Sousa, K., Mendonça, H., Lievyns, A., Vanderdonckt, J.: Getting users involved in aligning their needs with business processes models and systems. Bus. Process Manag. J. **17**(5), 748–786 (2011)

51. Leyer, M., Heckl, D., Moormann, J.: Process performance measurement. In: vom Brocke, J., Rosemann, M. (eds.) Handbook on Business Process Management 2, 2nd edn. Springer, Heidelberg (2015). https://doi.org/10.1007/978-3-642-45103-4_9

52. Dumas, M., La Rosa, M., Mendling, J., Reijers, H.A.: Introduction to business process management. In: Dumas, M., La Rosa, M., Mendling, J., Reijers, H.A. (eds.) Fundamentals of Business Process Management, pp. 1–31. Springer, Heidelberg (2013). https://doi.org/10.1007/978-3-642-33143-5_1

53. Albliwi, S., Antony, J., Abdul Halim Lim, S., van der Wiele, T.: Critical failure factors of Lean Six Sigma: a systematic literature review. Int. J. Qual. Reliab. Manag. **31**(9), 1012–1030 (2014)
54. Chakrabarty, A., Kay Chuan, T.: An exploratory qualitative and quantitative analysis of Six Sigma in service organizations in Singapore. Manag. Res. News **32**(7), 614–632 (2009)

Author Index

Printed in the United States
By Bookmasters